SpringerBriefs in Applied Sciences and Technology

SpringerBriefs present concise summaries of cutting-edge research and practical applications across a wide spectrum of fields. Featuring compact volumes of 50–125 pages, the series covers a range of content from professional to academic.

Typical publications can be:

- A timely report of state-of-the art methods
- An introduction to or a manual for the application of mathematical or computer techniques
- A bridge between new research results, as published in journal articles
- A snapshot of a hot or emerging topic
- An in-depth case study
- A presentation of core concepts that students must understand in order to make independent contributions

SpringerBriefs are characterized by fast, global electronic dissemination, standard publishing contracts, standardized manuscript preparation and formatting guidelines, and expedited production schedules.

On the one hand, **SpringerBriefs in Applied Sciences and Technology** are devoted to the publication of fundamentals and applications within the different classical engineering disciplines as well as in interdisciplinary fields that recently emerged between these areas. On the other hand, as the boundary separating fundamental research and applied technology is more and more dissolving, this series is particularly open to trans-disciplinary topics between fundamental science and engineering.

Indexed by EI-Compendex, SCOPUS and Springerlink.

More information about this series at http://www.springer.com/series/8884

Priyan Dias

Philosophy for Engineering

Practice, Context, Ethics, Models, Failure

Foreword by David Blockley

Springer

Priyan Dias
Department of Civil Engineering
University of Moratuwa
Moratuwa, Sri Lanka

ISSN 2191-530X ISSN 2191-5318 (electronic)
SpringerBriefs in Applied Sciences and Technology
ISBN 978-981-15-1270-4 ISBN 978-981-15-1271-1 (eBook)
https://doi.org/10.1007/978-981-15-1271-1

This Springer imprint is published by the registered company Springer Nature Singapore Pte Ltd.
The registered company address is: 152 Beach Road, #21-01/04 Gateway East, Singapore 189721, Singapore

To my wife Shanthi,
and our sons Ashwin and Sanjit

Foreword

As an engineer without a formal education in philosophy, Priyan Dias takes on a formidable task in writing a book on philosophy of engineering. He achieves his goal with great humility and awareness of the size of the challenge. He believes, and I agree with him, that most thinkers and practitioners see philosophy and engineering as independent topics, whereas they have much to offer each other. Since we first met in the early 1990s, Priyan and I have enjoyed many interesting discussions around these matters.

Professional philosophers have, until recently, largely regarded any philosophical issues of engineering as being subsumed under those of science. Practitioners tend to think of philosophical discussions as somewhere 'up-in-the-clouds' and almost totally irrelevant to the challenges of practice. As a result, philosophers have historically paid insufficient attention to the relationship between knowing and doing. Meanwhile, as science has become increasingly important to practice, engineers have allowed reflections and interpretations of what they do and how they do it to be undertaken by non-engineers. One consequence is that the voice of engineering has been weak. Critics have tended to focus on technology as applied science without recognising what Aristotle called *phronesis* or practical wisdom—the intellectual virtue of practical reasoning. Research indicates that most non-engineers associate engineering with construction, mechanics, building and fixing things but miss the aspects of design, innovation and creativity, problem-solving and aesthetics. Nobel was an engineer, and yet there is no Nobel Prize for engineering.

In this book, Prof. Dias poses some interesting questions. What defines engineering? Are engineers makers or thinkers? Is failure essential for success? What is the role of a model? Is engineering value neutral? In exploring the issues of practice, context, ethics, models and failure, he introduces us to four significant philosophers, Popper, Kuhn, Polanyi and Heidegger. Although they are not usually associated with engineering, Prof. Dias shows how each one has suggested ideas that shed light on the nature of engineering. Popper, for example, was a wide-ranging thinker, writing about science and society but hardly mentioned engineering. Nevertheless, his characterisation of evolutionary problem-solving is directly relevant to any form of practice. Kuhn likewise hardly refers to engineering, and his ideas are regarded by many professional philosophers as incompatible with Popper since they can seem to be a form of relativism, but herein lies the importance of theoretical models built for specific purposes. In science, purpose is a pursuit of absolute truth whereas, as Prof. Dias shows, the purpose of engineers is to fulfil a human need by making the best decisions that create safe artefacts. Central to those decisions is the availability of dependable theoretical models. Those models develop through processes like those identified by Kuhn but are clearly underpinned by ethical stances.

Professor Dias shows the critical role of ethics and value systems in engineering through Polanyi's three spheres of morality in science, the individual scientist, the scientific community and wider society. Faith and trust are pivotal values. Anyone doubting the importance of aesthetics as beauty in engineering should consider the awe-inspiring constructions of the pyramids, the beauty of Gothic cathedrals, the grandeur of a big bridge, the sleekness of an Aston Martin car or the sophistication of an iPhone. Engineers have a duty to care for our environment. Polanyi speaks of the freedom of the subjective person to do as he pleases being overruled by the freedom of the responsible person to do as he must. In the law of tort, this is a professional duty of care for others.

Heidegger may be the hardest philosopher in history to interpret since his writing is well known for being long, complex, dense and obtuse. Professor Dias largely avoids the famous 'isms' and labels of philosophers (such as phenomenology) as he points out how, at first, Heidegger affirmed traditional technology and the primacy of practice over theory but went on later to question more modern versions of science-based technology that is modern engineering. For example, Heidegger's concept of 'enframing' is not easy to grasp. Enframing is the way humans are said to be entrapped in processes of challenging nature that have consequent dangers. Indeed, Heidegger was beginning to articulate the need to erase the subject–object distinction. It is a division that has given us all the power and achievements of scientific reductionism while, at the same time in many people's view, it has taken away our awareness of the need to care for our planet. A wider and deeper understanding of enframing could be crucial as we together face the twenty-first-century challenges of climate change and computerisation.

Professor Dias admits there is much more to explore about the relationship between philosophy and engineering—such as uncertainty, sensitivity, hierarchy and emergence. He expresses the wish that this book will stimulate more thinkers to explore the issues and to inform the teaching of the philosophy of engineering within engineering programmes. In this, I think he succeeds and gives us much food for thought.

July 2019

David Blockley, DSc, FREng
Emeritus Professor of Civil Engineering
University of Bristol
Bristol, UK

Preface

Soon after graduating with a civil engineering degree in Sri Lanka, I came across a book in the university library written by David Blockley called *The Nature of Structural Design and Safety*. The author was a structural engineering academic, but approached his subject philosophically, in particular through the ideas of Karl Popper. I recall having fairly broad interests even in those days, but this was the first time I came across the idea that philosophy could be relevant to engineering. I pursued graduate studies at Imperial College London, focusing for my Ph.D. research on rather down to earth area of concrete technology. However, the much wider access to books and other resources helped me continue my interest in what would be called philosophy of science. In addition, Dr. Blockley gave a seminar at Imperial that I listened to very keenly.

These beginnings led me to spend a sabbatical year with Prof. Blockley on a Commonwealth Fellowship during the academic year 1992/93 in his system group at Bristol University. My association with David has continued since then, and I am grateful to his considerable contribution towards my thinking. Finding engineers with philosophical inclinations is not easy. But David and I have had conversations about philosophers ranging from Anaximander and Aristotle to Heidegger and Popper, and their implications for engineering. Many echoes of his ideas can be found in this book, which he kindly read and wrote a Foreword for.

My next sabbatical was spent at the Institute for Complex Engineered Systems at Carnegie Mellon University in 2000/01, where Prof. Susan Finger was my host and Dr. Easwaran Subramanian a colleague. Although the institute was largely concerned with hard engineering solutions to practical problems, it allowed leeway for my philosophical inquiry within an engineering culture—and in any case, I was on a Fulbright Fellowship that allowed me to pursue whatever I wished. The form of this book took shape during this sabbatical. I spent hours delving into the works of Karl Popper, Thomas Kuhn, Michael Polanyi and Martin Heidegger and the secondary literature on them. In addition, my intellect was sharpened through attending the lunch hour seminars at the Centre for Philosophy of Science housed in the nearby University of Pittsburgh, and finally being invited to speak there on 'Constructing a Philosophy of Engineering'. This also led to my introduction to

Gualtiero Piccinini, then a graduate student at that university. Gualtiero very kindly read and commented on everything I had then written about the four philosophers above—I owe him a large debt of gratitude. He is now a philosophy professor at the University of Missouri at St. Louis.

This book has been a while in the making. Much of the material I produced while at Carnegie Mellon has first seen the light of day as various journal papers. I am therefore extremely grateful to Loyola D'Silva of Springer Nature, Singapore, for agreeing to publish them as a collection. I also thank the publishers of the previous material for permission to publish it in this composite form—acknowledgements are made in each chapter. Amudha Vijayarangan and Raghavy Krishnan of Springer liaised with me during the production process. My thanks are also due to Dr. Chandana Kulasuriya, a former engineering graduate student of mine, who kindly read and commented on the entire manuscript. Chandana and I have had many philosophical discussions, some of which have been articulated in print. I thank Prof. David Elms and Dr. Bruce Vojak too, for reading and commenting on the book.

All my academic pursuits have been carried out as a faculty member at the University of Moratuwa, Sri Lanka, where I have enjoyed the customary academic freedoms for unfettered inquiry into whatever I have chosen to research. I am grateful for this and also for the opportunity to test my ideas by teaching students. In particular, I thank Prof. Asoka Karunananda for inviting me to teach a course on philosophy of science to graduate students following an M.Sc. program in Artificial Intelligence, where I have covered some of the material in this book.

I have benefitted from my association with Dr. Vinoth Ramachandra, an engineer like myself, with a Ph.D. in computational fluid dynamics also from Imperial College, but one who has made a complete career change by becoming an applied theologian, social analyst and acclaimed author while working in an international university student movement. He embodies the idea that we are capable of exercising our minds deeply in very different directions. I also acknowledge my interactions with Dr. Natasha McCarthy, who co-organized a 2008 workshop on Philosophy and Engineering at the Royal Academy of Engineering in London; and those with my fellow editors of *Civil Engineering and Environmental Systems*, Professors Paul Jowitt, Marc Maes and Mark Milke.

Finally, I thank Shanthi my wife for believing in whatever I have embarked upon; and also for her unreserved support, in spite of the many onerous responsibilities she herself has shouldered—this has included being principal of a leading girls' school in Colombo, a concert pianist and the linchpin of our family. We have two sons—one a practising structural engineer and the other a philosophically inclined lawyer. This book is dedicated to the three of them.

Moratuwa, Sri Lanka
August 2019

Prof. Priyan Dias
priyan@uom.lk

Contents

Chapter 1
Introduction: From Engineering to Philosophy

1.1 What Defines Engineering? Practice, Context, Ethics, Models, Failure

What comes to mind when we think of engineering? The subtitle of this book suggests that practice, context, ethics, models and failure are very important for engineering. Some of these will be obvious to the general public as well; for example, it is widely recognized that engineers are practical people. We sometimes talk about 'engineering a solution', which refers to solving a problematic situation, whether it involves people or places or both. Hence, engineering is also known as a '*problem solving* discipline'. In addition, the Latin root *genere* means to 'beget or produce', implying novelty too. This conveys the idea that engineering is an *opportunistic* discipline, which exploits a situation cleverly. Both problem solving and opportunistic acts require *ingenuity*, from the Latin *ingenium*, which can mean 'innate quality', 'intelligence' and 'talent'.

The idea of *practice* is often contrasted to *theory*; and the discipline of engineering probably displays this contrast more than any other. This is because engineering programs in universities are dominated by theory, while an engineering career will be largely practical. The tension between theory and practice is captured nowhere better than in the words of Donald Schon (1987, p. 3):

> In the varied topography of professional practice, there is a high hard ground overlooking a swamp. On the high ground, manageable problems lend themselves to solution through the application of research-based theory and technique. In the swampy lowland, messy, confusing problems defy technical solution. The irony of this situation is that the problems of the high ground tend to be relatively unimportant to individuals or society at large, however great their technical interest may be, while in the swamp lie the problems of greatest human concern. The practitioner is confronted with a choice. Shall he remain on the high ground where he can solve relatively unimportant problems according to prevailing standards of rigour, or shall he descend to the swamp of important problems where he cannot be rigorous in any way he knows how to describe?

So practice is arguably the most important philosophical issue where engineering is concerned. *Context* is perhaps the next. Engineering solutions can be very complex. According to Nobel Prize winner Herbert Simon (1996, p. 51), complexity is not a

P. Dias, *Philosophy for Engineering*, SpringerBriefs in Applied Sciences and Technology, https://doi.org/10.1007/978-981-15-1271-1_1

property of an entity itself, but rather embedded in the context (see quote below). Conversely, each differing context will require a different solution—whether for an urban transport system, a multi-storey building or a mobile phone.

> We watch an ant make his laborious way across a wind- and wave-molded beach. He moves ahead, angles to the right to ease his climb up a steep dunelet, detours around a pebble, stops for a moment to exchange information with a compatriot … [I]t is a sequence of irregular, angular segments – not quite a random walk, for it has an underlying sense of direction, of aiming toward a goal … [H]e has a general sense of where home lies, but he cannot foresee all the obstacles between. He must adapt his course repeatedly to the difficulties he encounters and often detour uncrossable barriers. His horizons are very close, so that he deals with each obstacle as he comes to it … [V]iewed as a geometric figure, the ant's path is irregular, complex, hard to describe. But its complexity is really a complexity in the surface of the beach, not a complexity in the ant. (Excerpt by Doyle and Marsh 2013)

Another way to appreciate the centrality of context for engineering is to look at the engineering design process. Steven Goldman (2017, p. 23) says that the crucial difference between scientific and engineering reasoning is that the former is characterized by *necessity* (since the laws of nature cannot be changed but only found) and the latter by *contingency* (since engineering solutions are dependent on and tailored to their contexts):

> The distinctiveness of engineering *vis-a-vis* science is most clearly revealed in the design process, understanding 'design' as that facet of engineering practice that produces a specification of the terms of a problem together with criteria of *acceptable* solutions to that problem. Both of these require multiple complex value judgements centred on what someone wants the outcome of the design process to *do*. The outcome of the design process is not the revelation of a pre-existing state of affairs as in science, but an act of creation. What engineers enable is an outcome determined by the wilfulness motivating a desire for what some outcome must do and how it needs to do it. The problems that engage scientists are 'there', waiting to be recognized. Engineering problems, by contrast, are created by people who *want* to do something specific and are constrained in various ways, to a degree by what nature will allow, but primarily by highly contingent factors that, from a logical as well as a natural perspective, are arbitrary: time, money, markets, vested interests and social, political and personal values. (Italics and quotation marks from original)

Engineering impacts people, generally in large numbers; and also our environment, often for long durations and in irreversible ways. So *ethics* is imperative for engineering, but there are two ways in which they get divorced from each other. The first is to see the latter as purely mechanical as opposed to moral as well (Stretch 1986). The second is to treat engineering and technology as being essentially *neutral*. Ethics (i.e. how we should conduct ourselves) is treated very seriously in all philosophical traditions; and because ethics is rather neglected within engineering (unlike for example in medicine), it becomes an important philosophical issue.

Most engineers deal with *models*, where the artefact to be constructed or fabricated is represented in some way. It has been said that humans are perhaps the only animals that do most things twice—first in our minds and then in the real world. This would apply to the engineering task of fabricating or producing something too. When our artefacts were very simple, a correspondingly simple mental image or model may have sufficed. But with increasing complexity, the models had to be drawn in sketches or formed in scaled versions of the artefact. At present, most engineers would understand modelling to be mathematical in nature, and implemented in computer software. Otherwise it would be impossible to predict how a supertall building responds to an earthquake, or an ultra-tiny microchip emanates undesirable heat.

The question of whether the real world can be captured in a representation is however another deeply philosophical question that engineers would be wise to consider. George Box (1976), a British statistician, is credited with the sentiment that: "All models are wrong, but some are useful". This is a sentiment that most engineers will identify with. Albert Einstein is supposed to have said: "Not everything that can be counted counts, and not everything that counts can be counted." Engineers would agree with this too; their problem however is how to incorporate in their models the things that count but cannot be counted.

We come finally to the issue of *failure*. This may seem counter-intuitive because engineers are looking to create things that *don't* fail. But Henry Petroski has said: "Failure is central to engineering. Every single calculation that an engineer makes is a failure calculation. Successful engineering is all about understanding how things break or fail". His first and probably best book is titled *To Engineer is Human: The Role of Failure in Successful Design* (Petroski 1985).

These then are the issues considered in this book—practice, context, ethics, models and failure. They apply to all branches of engineering, and in some ways to all human endeavour. Because while humans are commonly known as *homo sapiens*—beings with intellectual capacity, they are also distinguished from other species by being *homo faber*—beings who make things; and the only species that creates their own environments rather than merely adapting to them. The issues may not be exhaustive, and other approaches may prioritize different ones, but these are a good set to start with.

1.2 Which Philosophers Do We Turn to? Popper, Kuhn, Polanyi, Heidegger

It so happens that Karl Popper, Thomas Kuhn, Michael Polanyi and Martin Heidegger, the four philosophers we consider in this book, tackle the issues of practice, context, ethics, models and failure in specific ways that are very relevant to engineering. Heidegger's main book is called *Being and Time*, where the notion of 'being' refers to the 'human way of being', which he calls 'being in the world'—in other words, humans are rooted in their *context*. The key example he uses to describe 'being' is that of a carpenter engaged in his *practice*.

Polanyi, whose approach to philosophy is very different to Heidegger's, nevertheless highlights the importance of *practitioner* involvement in the pursuit of science (e.g. through both the passion and the discretionary judgement required for research), although it is supposedly characterized by 'practitioner detachment'. His main book is titled *Personal Knowledge* (Polanyi 1958). Both Heidegger and Polanyi deal with *ethics* too—the former through advocating a suspicion of technology, and warning against treating it as merely neutral; and the latter by saying that a scientist's personal freedom to exercise judgement needs to be balanced by a self-imposed constraint to be responsible.

Kuhn's main contribution to our inquiry comes from his view that scientific theories are constructed by humans, and not necessarily a reflection of what is true of the world. While most scientists may not be happy with this, engineers would readily accept it of their *models*, which are only required to be useful or dependable for fabricating artefacts. In fact, engineers represent the world in different ways; which could range from practical 'rules of thumb' derived over many years of practice, to highly theoretical mathematical models that may still not fully capture aspects that 'count but cannot be counted'.

Finally, Popper presented a scientific methodology in which 'falsification', or proving a theory false, was the quickest way to the growth of scientific knowledge. This resonates very strongly with the notion of *failure*, whether as a basis for engineering design calculations, or as an important way in which engineering knowledge grows—i.e. by learning through failures of engineering artefacts. Popper's scientific methodology was also called a problem solving approach, once again making him very relevant to engineering. One collection of his writings is titled *All Life is Problem Solving* (Popper 1999).

The above philosophers are almost iconic figures of 20th century philosophy. Popper, Kuhn and Polanyi can properly be called philosophers of science; they focused on *epistemology*, the branch of philosophy that deals with *knowledge* or *how we know*. Since engineering is at least partly based on science, that seems a good way to start. In addition, they have challenged the view that science is a 'cool, detached' discipline, since it also depends on human imagination (Popper), consensus (Kuhn) and judgement plus artistry (Polanyi)—factors that are very characteristic of engineering. Heidegger focused on *ontology*, the branch of philosophy that deals with *being* or *how we are*, which includes *action* or the way we *do* things; and since engineering is a very practical discipline, he becomes a very important philosopher for engineers. On the one hand, he was a critic of technology; on the other, he stressed the importance of 'doing' over 'thinking', thus endorsing the value of the very pragmatic engineering profession.

Other authors have also developed ideas for engineering using these philosophers. David Blockley's (1980) *The Nature of Structural Design and Safety* borrows many ideas from Popper. William Addis' (1990) *Structural Engineering: The Nature of Theory and Design* is largely based on Kuhn's concepts of scientific revolutions and paradigm shifts. *Understanding Computers and Cognition: A New Foundation for Design* by Winograd and Flores (1986) draws significantly from Heidegger's notions of 'embeddedness' and 'breakdowns' for their suggested approaches to design. Such precedents give validity for our own use of these philosophers.

Our approach, of using insights from established philosophers to bear on engineering, is why this book is titled 'Philosophy *for* Engineering' rather than 'Philosophy *of* Engineering'. We have not used philosophers of engineering as key sources, because that field is emerging only now. There are many ways to approach a philosophy of engineering, and none of them have become strongly established. Two philosophers and their major books worth mentioning are Carl Mitcham's (1994) *Thinking Through Technology: The Path between Engineering and Philosophy*, and Michael Davis' (1998) *Thinking like an Engineer: Studies in the Ethics of a Profession*. McCarthy's (2009) *Engineering: A Beginner's Guide* is also a highly accessible introduction. Some others, written by engineers, are Vincenti's (1990) *What Engineers Know and How They Know it*; Koen's (1985) *Definition of the Engineering Method*; and Florman's (1994) *The Existential Pleasures of Engineering*.

1.3 How Is This Book Engineered?

This chapter presents the approach to the book. It identifies the issues we shall tackle and the philosophers we turn to, establishing the linkages between the two. Before we get into either the issues or philosophers however, we present in Chap. 2 some of the tensions felt by engineers with respect to their influence, role and knowledge. We show that these map onto ethics, ontology and epistemology respectively, which are major areas of philosophy.

Chapters 3–8 form the meat of the book. They can be read and profited by as separate chapters, but there is a progression too. Chapter 3 deals with Popper's relevance to the issue of *failure* in engineering—the way we should use it in both design and diagnosis. Chapter 4 goes on to consider what we can learn from Kuhn regarding engineering *models*. Popper and Kuhn are philosophers of science who are relatively easy for scientists and engineers to appreciate. We start with them for that reason, but also because the notions of failure and models are ones that engineers come across regularly if not daily. Chapters 3 and 4 however are somewhat long.

Chapters 5 is about *ethics* but also about aesthetics, two areas that are loosely described as being 'subjective'. We show how they can in fact be seen as 'inter-subjective', where significant agreement among individuals is possible, resulting in shared values. Polanyi is the philosopher we draw on in this chapter—personal knowledge arrived at with universal intent is his *mantra*.

Chapters 6 and 7 are on Heidegger, with the first of these continuing the discussion on ethics, while the second deals with the rather difficult notion of 'being', from which we develop ideas about an engineer's *practice* in *context*. Heidegger is a difficult philosopher to understand, but his ideas are both relevant to technology and resonate with an engineering outlook. It is hoped that our exposition of Heidegger will introduce him to a wide engineering community.

Chapter 8 brings together both Polanyi and Heidegger as providing a basis for *practice based knowledge*, something that is flagged as an area of need in Chap. 2. We show how they have complementary views on the subject; and also suggest that Artificial Intelligence could be a tool for providing some techniques for it. We close with Chap. 9, which draws out commonalities in our four philosophers with respect to practice, context, ethics, models and failure.

We do not merely use 'cherry-picked' ideas from the four philosophers to support our key engineering concepts. Instead, a genuine summary of their most important ideas are presented. Engineers may find this somewhat challenging, but will appreciate the engineering relevance of those ideas. Similarly, philosophers may find the philosophy somewhat elementary, but have their interest aroused by our applications. If we succeed in demonstrating that core engineering concepts are supported by philosophical ideas, it will hopefully make engineering more erudite and philosophy more useful.

Since this is a book about philosophy for engineering, engineering examples have been used to flesh out some of the ideas. Most if not all the examples are from the discipline of structural engineering. Apart from this being my own discipline and the one I am most familiar with, it is a field that all of us encounter frequently, since we live most of our lives within structures that encompass us—even when flying! Structures are also probably the most prominent and visible embodiments of our civilization as humans—most tourist destinations feature structures from their history. This does not make the book irrelevant for other fields of engineering—the examples are specific but the ideas universal for all forms of engineering. As humans with minds and bodies we are quite adept at going back and forth between universal mental concepts and specific physical examples or instances.

Finally, much of this book presents engineering as being *distinct* from science; this is largely to correct views that engineering is 'merely applied science' or 'an approximation of science' or even 'inferior to science'. On the other hand, a careful reading will reveal that the book often emphasizes the *centrality* of science to engineering—in fact its main sources are philosophers of *science*; and also the *complementarity* of science and engineering approaches to problem solving. We do argue however that engineering both encompasses and is more sophisticated than science.

1.4 Summary

- Engineering can be seen as characterized by practice, context, ethics, models and failure. Many authors have highlighted the importance of these features for engineering, which is both a problem solving and opportunistic activity. Science is central to engineering, but engineering is distinct from and encompasses science in many ways.
- Karl Popper, Thomas Kuhn, Michael Polanyi and Martin Heidegger are iconic 20th century philosophers who have dealt with the above ideas largely in the context of science, but in ways that are highly relevant to engineering. Hence we use them to reflect philosophically on engineering. Other engineering authors have also appropriated these philosophers for similar purposes, thus validating our approach.

References

W. Addis, *Structural Engineering: The Nature of Theory and Design* (Ellis Horwood, New York, 1990)

D.I. Blockley, *The Nature of Structural Design and Safety* (Ellis Horwood, Chichester, 1980)

G.E.P. Box, Science and statistics. J. Am. Stat. Assoc. **71**, 791–799 (1976)

M. Davis, *Thinking Like an Engineer: Studies in the Ethics of a Profession* (Oxford University Press, New York, 1998)

M. Doyle, M. Marsh, Stigmergy 3.0: From ants to economies. Cogn. Syst. Res. **21**, 1–6 (2013)

S.C. Florman, *The Existential Pleasures of Engineering*, 2nd edn. (St. Martin's Press, New York, 1994)

S.L. Goldman, Compromised exactness and the rationality of engineering (Chap. 1), in *Social Systems Engineering; The Design of Complexity*, ed. by C. Garcia-Diaz, C. Olaya (Wiley, Oxford, 2017), pp. 13–29

B.V. Koen, *Definition of the Engineering Method* (American Society for Engineering Education, Washington, DC, 1985)

N. McCarthy, *Engineering: A Beginner's Guide* (Oneworld Publications, London, 2009)

C. Mitcham, *Thinking Through Technology: The Path Between Engineering and Philosophy* (Chicago University Press, Chicago, 1994)

H. Petroski, *To Engineer is Human: The Role of Failure in Successful Design* (St. Martin's Press, New York, 1985)

M. Polanyi, *Personal Knowledge: Towards a Post-critical Philosophy* (University of Chicago Press, Chicago, 1958)

K.R. Popper, *All Life is Problem Solving* (Routledge, London, 1999)

D.A. Schon, *Educating the Reflective Practitioner* (Jossey-Bass, San Francisco, 1987)

H.A. Simon, *The Sciences of the Artificial*, 3rd edn. (MIT Press, Cambridge, 1996)

L. Stretch, *Engineering: Mechanical or Moral Science?* (Becket Publications, Oxford, 1986)

W.G. Vincenti, *What Engineers Know and How They Know It: Analytical Studies from Aeronautical History* (Johns Hopkins, Baltimore, 1990)

T. Winograd, F. Flores, *Understanding Computers and Cognition: A New Foundation for Design* (Ablex, Norwood, 1986)

References

[illegible]

Chapter 2
Are Engineers Makers or Thinkers?

2.1 Do Engineers Have an Identity Crisis?

The notion of 'identity' is in vogue today. In some cases it has to do with individual identity: captured in the question "Who am I?" In others it has to do with group identity—references are made to 'post-colonial identity' or 'African-American identity'. We have an identity crisis when we are not sure about who we are or what we are doing. Thoughtful or reflective engineers could be suffering from an identity crisis regarding their collective engineering identity; and there are at least three reasons why (Dias 2013).

First, there is a crisis regarding the engineer's *influence*. There was a time when engineering was synonymous with the progress and uplifting of humanity. The great 19th century bridge builder Isambard Kingdom Brunel was second only to Winston Churchill in a poll, held at the turn of the millennium, to determine the greatest Briton of all time. Today however, we have an environmental crisis brought about by rapid industrialization; and a society that has become 'individualized' because people are more focused on their 'technological gadgets' (e.g. 'smart phones') than on relating to each other. Since engineers are at the forefront of industry and technology, there arises the question as to whether *engineers are doing more harm than good* through their actions. The study of such actions, their motivations, and impacts is that branch of philosophy called *ethics*.

Next, there is a crisis regarding the engineer's *role*. Most students who enroll in engineering undergraduate programs have a strong background and interest in science. They are good at analysis. Practising engineers on the other hand have to produce something or make things happen. That involves combining products and processes; getting people to either work with you or accept what you are doing; and achieving all this within a limited budget and time frame. In other words, they must be good at management and synthesis. The question then arises as to whether *engineers are scientists or managers*. Genuine scientists and capable managers are both valued in most societies, but engineers run the risk of becoming neither in trying to be both. In some cases engineering graduates move on to become managers; while

P. Dias, *Philosophy for Engineering*, SpringerBriefs in Applied
Sciences and Technology, https://doi.org/10.1007/978-981-15-1271-1_2

others opt for academic careers as engineering scientists. The study of roles within the wider study of 'being' is that branch of philosophy called *ontology*.

Finally, there is a crisis regarding engineering *knowledge*, which overlaps the one regarding role. Most university programs in engineering are filled with theoretical subjects that are largely 'mathematics in disguise'. Engineering practice on the other hand is predominantly practical in nature, with great reliance placed on established procedures (or 'rules of thumb'), specified guidelines (or 'codes of practice'), and that indefinable element called 'engineering judgement'. Therefore, we can ask whether *engineering knowledge is theoretical or practical.* In some situations, engineers have difficulty in explaining how their knowledge differs from that of a technician or even craftsman, because of this reliance on rules of thumb. The study of knowledge is that branch of philosophy called *epistemology*.

The above questions are valid for engineers in most if not all societies. It is the duality ('or') posed in the questions that creates the discomfort or *angst*—a German word that is loosely understood as *anxiety* in English. This may not be the everyday experience of engineers, but if they are not confident about the answers to these questions, they could experience doubts about their self-worth or social value—and hence their identity crisis. Ethics, ontology and epistemology are well established branches of philosophy. So the engineer's identity crisis is a philosophical one, and highlights the importance of philosophy for engineering.

Usually however, technology and philosophy are seen as being poles apart (see Fig. 2.1), the latter focused on *thinking* and the former on *doing*. This tension is conveyed by the double-headed horizontal arrows in Fig. 2.1. This figure shows other entities that flow from technology and philosophy (in single headed vertical arrows), and there could be such tensions at these levels too. For example, while engineering is considered by some to be 'nothing but applied science', others would say that science is merely one of the tools at the disposal of a sophisticated engineer. While some would say that practice is based on theory, others may say that it is practice that generates the very concepts required for theory. Figure 2.1 cannot capture all

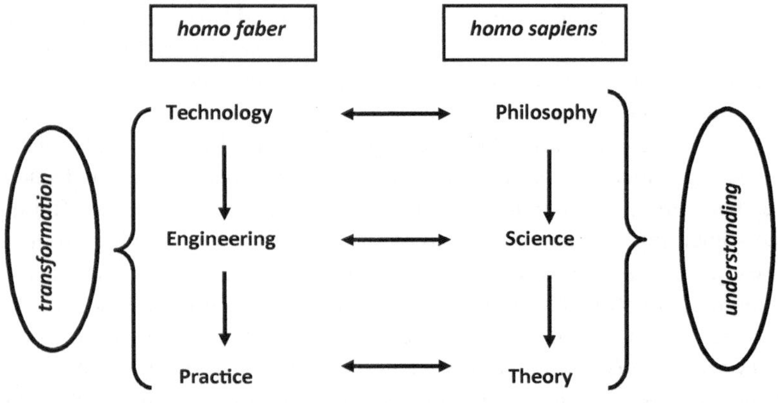

Fig. 2.1 Transformation versus Understanding (from Dias 2013)

the intricate relationships between the entities portrayed, but it serves to guide the discussion in this chapter. It is interesting that the entities on the right hand side are concerned with *understanding* (which is the goal of *homo sapiens* or ‘the wise man’), while those on the left hand side with *transformation* (which is the goal of *homo faber* or ‘man the maker’). Recall Karl Marx’s comment, i.e. “philosophers have tried to understand the world; the point however, is to change it”. All three identity crisis questions raised above can be related to the question of whether an engineer is a maker (*homo faber*) or a thinker (*homo sapiens*).

2.2 The Engineer’s Influence: More Harm Than Good?

We answer this question in the context of the tension between philosophy and technology, given that engineers are the main agents of technology. Some 20th century philosophers (Ellul 1948; Heidegger 1977) have charged technology with being a harmful influence, quite in contrast to the ‘humanizing’ influence of philosophy and other liberal arts. The ill effects of technology can be categorized into at least four aspects (Dias 2003). The most obvious is the hazardous nature of some technologies, the prime example being nuclear technology. In addition, technology can promote injustice, for instance through infrastructure projects where social costs are borne by the poor and the benefits reaped by the rich. Technology can have adverse sociological impacts too—consider the way in which visual screens (whether televisions or computers) tend to destroy family conversation and interaction. Finally, and most subtly, it can have undesirable psychological impacts. Has technology created a society where ‘technique’ is all important, as opposed to understanding (of phenomena) or even genuineness (in relationships), reflected in the growing number of ‘how to’ books? Heidegger (1977) claims that man has been ‘enframed’ by his own technology. We shall consider more of this in Chap. 6.

The American engineer Samuel Florman (1994) however refutes these charges, and also points to the benefits bestowed upon the world by technology, in areas such as transportation and health, and by general improvements in the standard of living. In other words, ‘humanization’ can be seen, not so much as an enjoyment of the arts (which in most societies is enjoyed only by a relatively few), but rather as the liberation from ‘slavery’ brought about through technology, as described below by Karl Popper (1999, p. 104), himself a philosopher of science who we shall meet in Chap. 3. We must remember while reading this however that many parts of the world have not yet been liberated from such drudgery.

> Perhaps even more important, morally, was the great liberation of domestic slaves (also known as maids), which became possible largely through household mechanization. This tremendous revolution, and the emancipation that all but the very richest women experienced at that time, is today remarkably little remembered, even though it was a liberation from heart-rending slavery. Who today has any idea what it meant when all water had to be fetched and carried, when coal had to be brought in for heating, when all washing had to be done by hand, and when there were still oil lamps with wicks?

Florman (1994) also argues that the engineer's activity of making things and engaging in work is a way of experiencing his humanness, relating to the earth and producing what he calls 'existential joy'. He does admit however, that the work of engineers may lead to unforeseen negative consequences, but applauds them for trying to improve the world. Florman asks engineers to take courage from Sisyphus, the character from Greek mythology who was condemned to keep rolling a stone up a hill, only to have it falling back as he approached the summit. Florman sees Sisyphus as heroic—someone who refuses to give up even though his work is undone from time to time. In this context, it is interesting to note that modern initiatives against some of the ill-effects of technology want to use technology itself to cure those ills (Feenberg 1999). For example, technology intensive underground sequestering of carbon dioxide is being considered for reducing the consequences of burning fossil fuels; and it is the 'high-tech' internet that is used for getting greater access to knowledge and to communicate, by people who feel they are marginalized and alienated by technocrats.

The anti-technology attitude does not arise only because of technology's rather recent negative effects. For many centuries university education was seen primarily as a process for creating better human beings, through the dissemination and discovery of knowledge that was not directly 'useful'. In Ancient Greece for instance, 'pure speculation' (by 'thinkers') was considered to be a more noble activity than the doing or making of useful things (by 'workers'). Although Aristotle (2000) recognized and appreciated the importance of practical wisdom for action (*phronesis*), he concludes nevertheless by giving pride of place to theoretical wisdom (*sophia*). Consider also the following description of Archimedes given by Plutarch (Blockley 1981):

> Yet Archimedes possessed so high a spirit, so profound a soul, and such treasures of scientific knowledge, that though these inventions had now obtained him the renown of more than human sagacity, he would not deign to leave behind him any commentary or writing on such subjects; but, repudiating as sordid and ignoble the whole trade of engineering, and every sort of art that lends itself to mere use and profit, he placed his whole affection and ambition in those purer speculations where there can be no reference to the vulgar needs of life.

Florman (1994) says that this mind-set, together with the Biblical New Testament emphasis on the spiritual as opposed to the material, has given technology a bad image or low status in western culture. In eastern cultures too, the role of the sage (or 'guru') has been exalted over that of the worker, with the strict caste system in India, for example, perpetuating social barriers for generations. Blockley (1981) says "We must break the chains of Ancient Greece"; but how? Florman (1994) suggests, as least for western culture, that engineers dig deeper into their heritage for evidence that 'making' is indeed a noble pursuit, e.g., into the Old Testament, where the ability to perform various skilled crafts is ascribed to the indwelling of the Spirit of God; also into the pre-Socratic era, where craftsmanship was held in high esteem by Homer, who gives great technical detail regarding the making of Odysseus' raft and Achilles' shield, covering both tools and materials.

Meanwhile, a vast expansion of university education has taken place in the past century; with research and innovation directed towards wealth creation and problem solving. Science and engineering faculties were well funded; while humanities faculties, other than in the most prestigious universities, experienced gradual decline in both funding levels and student numbers. Technologists now use the term 'soft' to undervalue what goes on in arts faculties; while those in the humanities complain that human values are submerged by a technological mindset. At the end of the day however, in many societies an 'educated' person (or 'intellectual') is considered to be one who has knowledge of literature and culture, rather than one who can describe an internal combustion engine or an integrated circuit. As we shall see later on (Chaps. 6 and 7), Martin Heidegger is probably a good 'patron' philosopher for engineers. On the one hand, his view of what he called the 'human way of being' could be called an 'instrumentalist' one—we 'are' and 'do' before we 'think' (Dias 2006). On the other, he was suspicious of technology where it destroyed the rich diversity of the world and human interaction with it (Dias 2003).

So, are engineers doing more harm than good? Whatever accusations are made against engineers, those who level such charges would probably not want to live in a world without technology and engineering influence. Where the capacity for humanization is concerned, technology has credentials that can rival those of philosophy, as articulated above by Popper. Furthermore, as emphasized by Florman, engineers can be proud that they are men and women of action—being *makers* in other words—rather than merely engaging in 'pure speculation'. Engineers may feel inferior about their intellectual status however, i.e. their place on the scale of *thinkers*, and we deal with this at the next level of the framework (Fig. 2.1), which considers the tensions between engineering (very much a part of technology) and science (a development of philosophy).

2.3 The Engineer's Role: Scientist or Manager?

In order to shed light on the role of an engineer, we consider engineering design, which is a good reflection of engineering practice as a whole. We could view (engineering) science as the core or kernel of engineering design knowledge; a core however that is contained within outer layers of knowledge such as engineering idealization, margins of safety, design philosophy, design context and engineering process—see Fig. 2.2. Before using engineering science theories, we have to adopt a particular design philosophy, decide on margins of safety, and idealize the real world into a model to which scientific or mathematical theories can be applied; this has to be done within a design context that would be defined by the location and people (clients, customers) concerned. All of the above is enveloped in an engineering process that will involve collaboration and communication, not least with those who will fabricate, maintain and use the designed artefact.

Let us use the simple structural engineering assembly of a reinforced concrete beam supported between two columns in Fig. 2.3 as an example. It is the idealized beam that is analyzed using engineering science (to find, say, the design bending moment at midspan). However, is the beam best idealized as a simply supported

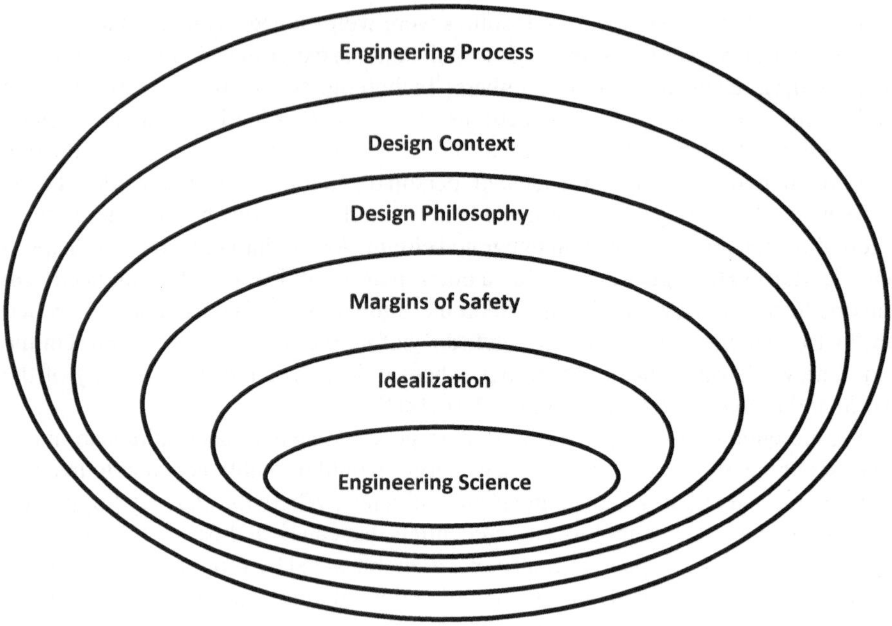

Fig. 2.2 Engineering design knowledge (after Dias 2013)

Fig. 2.3 Idealization of a real structural element (from Dias 2013)

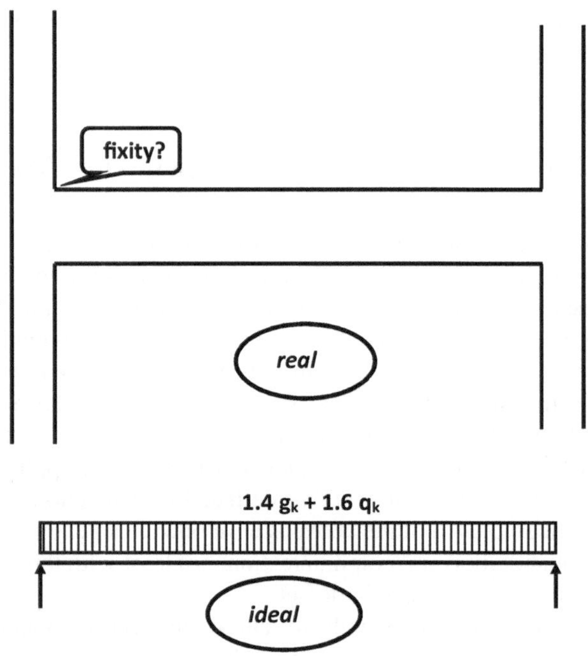

beam (free to rotate on the column supports at its ends) or as a fixed ended beam (where rotation at supports is prevented by fixity to the columns)? The former is true in most situations, since the beams are stiffer than the columns. We also need to apply appropriate safety factors to the load that is anticipated. For example, we assume in most cases that the loads are distributed uniformly. However, the expected loads are increased by factors, in order to ensure a margin of safety; in this example the 'permanent' loads are factored by 1.4 and the 'imposed' ones by 1.6. We also have to adopt some design philosophy to make allowance for the restraining moments at the column supports, where the fixity is not known precisely. Another established structural engineering design philosophy is that structural elements are designed to be 'ductile'—in other words, while they are designed not to fail under the anticipated loading, if they do (say due to overloading), they should fail in a way that is gradual and ductile (as opposed to sudden and brittle). Blockley and Henderson (1980) use the term "calculation procedure model" to describe this entire process; it is not confined to calculations alone, but incorporates all the other decision making procedures.

All of the above occurs within a design context too. There is a 'narrow' design context, which could refer to the advantages and drawbacks of say the proposed construction site itself. For example, even if the column size required in Fig. 2.3 is smaller than that of the beam, we would make it larger in an earthquake zone, because we want to avoid the catastrophic consequences of failure in the columns. Broader than this however, are the ever present contexts of time, cost and politics within which all engineering has to be carried out. This context dependency or *contingency* means that engineering solutions will never be exact. As Goldman (2017) says, the rationality of engineering is one of 'compromised exactness'—the compromises being due to the overall process and especially the context. He presents such a rationality as being highly sophisticated and eminently suitable for social and economic decision making as well.

We can finally think about the process through which everything above has to be carried out. Within an engineering design office there will be senior engineers supervising younger colleagues. The latter will probably be responsible for carrying out the detailed calculations; or using computers for that purpose. However, the senior engineers will need to check the work of their juniors without spending too much of their precious time; they will need to decide what to check and where, in the structure or other artefact. Process also involves liaising with other professionals, which for structural engineers would mean architects and services engineers; and also the construction engineers. Such liaising could involve making compromises in their design solutions, or being able to negotiate for their own preferences by using persuasive arguments. There is also the notion of design for constructability—e.g. when specifying the steel reinforcement bars for the elements in Fig. 2.3, we should ensure that the sizes and spacing chosen allow concrete to be vibrated through them.

The message of Fig. 2.2 is that *engineering is broader and richer that science*. This breadth and richness create complexity that has to be managed for practical problem solving. It should be noted that the term 'complex' is used to denote richness in structure (whereas the term 'complicated' denotes abundance of detail). It is this structural richness that constitutes the intellectual challenge of engineering.

Another complexity is the uncertainty associated with engineering; this uncertainty has been classified (Blockley and Godfrey 2000) as Fuzziness, Incompleteness and Randomness (FIR). *Randomness* describes the variations to be expected in loading and material strength; we typically use statistics to tackle it. In order to be safe, we try to use in our calculations material strengths *below* which not more than 5% are expected, and loads *above* which not more than 5% have been known to occur in the past. *Fuzziness* relates to the imprecision in assigning states to an entity—for example, the judgement as to whether the beam in Fig. 2.3 is fixed ended or simply supported (hinged). Attempts to deal with fuzziness range from fuzzy set theory (Blockley 1980) to simple approaches such as designing for the worst effects of both fixed and hinged end conditions. *Incompleteness* has to do with the lack of knowledge about possible future scenarios, for instance the question as to whether the beam will be overloaded at any time during its design life, and by how much. Design approaches such as using the 'worst credible load' or designing for ductility (so that there is warning even if failure occurs) can be used. Interval probability theory is a more sophisticated approach (Blockley 2013).

Yet another type of complexity is that engineering problem solving often requires *abductive* reasoning, where a cause has to be posited, given an effect (observed or desired) and a known set of rules (Dias 2010). For engineering design this means that a solution (of which there could be many) has to be proposed for a desired performance, given the established rules of structural mechanics (and other associated heuristics). The fact that there will inevitably be more than one solution that can deliver the required performance is what constitutes the challenge. This is also called the solution to an 'inverse problem'. Apart from all of this, the 'calculation procedure model' has to make allowance for human error (or even malice) and accidents as well. So the engineering role is that of managing a process that involves people, procedures and products to deliver quality, including both safety and economy.

So, are engineers managers or scientists? From the above discussion we must conclude that engineers act more like holistic managers than specialized scientists, although their practice is grounded in science. This emphasizes yet again that the engineer is a *maker*—not only in the narrow sense of making things, but also in the broader sense of 'making it happen'. However, the complexity that (s)he has to tackle requires a particular kind of knowledge, understanding and even wisdom, thus making it appropriate for the engineer to be very high on the scale of a *thinker* too. We are now ready to consider the tensions between practice and theory, the third level in the framework of Fig. 2.1.

2.4 The Engineer's Knowledge: Theoretical or Practical?

There are many dimensions to the theory versus practice debate. As stated at the start of this chapter, most engineering programs are dominated by theoretical subjects, not only to ground engineering students in the 'kernel' of science (Fig. 2.2); but also to justify the existence of engineering as a university discipline, which has to be characterized by a body of theoretical knowledge. Engineering graduates discover however, that 'rules of thumb', 'codes of practice' and 'engineering judgement' dominate the actual practice of engineering. We can say therefore that engineering knowledge is largely practical, although it has to be based on theory.

Heidegger's (1962) example of a carpenter hammering a nail is very insightful for resolving this practice-theory tension. The 'primordial' (or immediate) experience of the carpenter is a seamless web of activity without any deliberate rationality on his part. He just picks up a hammer and drives a nail into a wall without analyzing what he is doing; he is practiced at doing so, being something natural for him. However, when there is a 'breakdown' in this 'everyday' experience, say when the hammer is too heavy, the carpenter will have to resort to 'mentality' and consider properties such as the weight of the hammer object; the notion of 'weight' becomes important. Again, if the head comes off the handle, once again he will have to give careful attention to solve the problem, and concepts such as 'jointing' will surface in his mind. In fact, Heidegger considered that scientific observation and reflection originated from such breakdowns in everyday activity or practice. So it is practice that gives rise to theory and not the other way around.

However, this illustration also underlines the necessity of theoretical training for engineers. Although they may be using mostly practical intelligence (e.g. 'rules of thumb') in their routine work, they will need a bedrock of theoretical knowledge to fall back on when faced with problems that intrude into their practice. Many professional engineering organizations, in the process of admitting engineers to full membership after a period of work-based training, are interested in finding out about problems encountered during the engineer's work, and how engineering 'first principles' were used to overcome them (Dias 2006). Heidegger's carpenter is a very important 'parable' for engineers. We return to it in Chap. 7.

There are other aspects of the interaction between theory and practice also worth looking at. For example, most engineering academics would say that the 'theory' components of a course should be taught before introducing students to 'practical applications'. In an overall sense, an engineering graduate would be seen as putting into practice the theory learnt at university. However, Patrick Nuttgens (1980), an architecture professor at York University who became the founding director of Leeds Polytechnic in the U.K. in the early 1970s, argues that children first learn about the world by practice before they acquire a theoretical framework; and that technical education should reflect this. Some 'sandwich' courses in engineering programs embody this to an extent. Students are exposed to practical experience during their period of study; and in some rare cases at the start of it.

Also, practice itself is now considered to be a rich source for theory, especially theories regarding engineering design; and the process of engineering design has been equated to theory building (Monarch et al. 1997). There are echoes here of 'grounded theory' (Glaser and Strauss 1967) used mostly by sociologists, where 'theory' is derived from the analysis of documents and transcripts of unstructured interviews. A broad philosophy of practice has been promoted for some time now (e.g. Goranzon 1995), with contributions from philosophers, engineers, craftsmen and actors; parallels have been drawn between actors and engineers. The attempt is to show that knowledge is very often acquired from practice (perhaps under apprenticeship), rather than from theory alone. Donald Schon (1983) wrote a very influential book called *The Reflective Practitioner*, which was subtitled *How professionals think in action*. The main theme of the book is that 'reflective practice', i.e. reflection on one's professional practice, generates practice based knowledge that is invaluable and very different from the theoretical knowledge that is embedded in 'technical rationality'. His ideas have been applied to engineering in general (Blockley 1992) and engineering design in particular (Dias and Blockley 1995; Dias 2002), for which it is argued that a combination of reflective practice and technical rationality is required.

We conclude this section by affirming that the knowledge used by engineers during their professional careers is mostly practical in nature, once again reinforcing the *'maker'* image of the engineer. We have shown however, that there is considerable interplay between theory and practice, and that many recent initiatives promote the intellectual status of practice. Despite this new focus on practice however, the greatest 'shortcoming' in practice-based knowledge is its lack of formalism. It is theoretical formalism that gives science its credibility and prestige in the academy and even in wider society. The practical knowledge of engineers is often perceived as 'just common sense'. In fact, even craftsmen and technicians are seen as having such knowledge, so that it is not valued as being intellectual. The place of the engineer among the ranks of *thinkers* is thus challenged. This challenge can be met through efforts to formalize practice.

2.5 Formalizing Practice

The formalization of engineering practice will strengthen the engineer's image as a *thinker*, while reinforcing his position as a *maker*—an agent of transformation. There are two levels at which formalization needs to evolve—at the conceptual level that deals with the engineering approach, and at the technical level that deals with practice-based knowledge. Systems approaches can be seen as providing a formalization at the conceptual level (Dias 2008). Formalization at this level is not easy, as best expressed by David Elms (2010):

> The systems approach is not easily systematised, so to speak, partly because of the breadth of the issues involved, but more generally because there is no narrow set of applications allowing development of an easily focused theory. Structural analysis, for example, has techniques fine-tuned to dealing with structures, but the systems approach can be applied to anything. It has no natural boundaries. What is needed is not so much a set of immediate techniques as general principles and overarching concepts for giving the approach its power and its constraints......The trap...to avoid [is] being so general as to be ineffective, hence specific guides and ideas are needed.

There have been many such frameworks proposed in the literature that interested readers can pursue. The reflective practice loop is one of them, consisting of the components reflection—action—world—perception—reflection (Blockley 1992). A development of this is the design—build—operate loops in the three spheres of purpose, process and people (Blockley 2010a). Senge (1992) has demonstrated that a number of management scenarios can be modeled with three basic elements—namely a reinforcing loop, balancing loop and process delay. Blockley (2010b) has proposed a framework that he calls 'new process', where a process is seen as a relationship between sub-processes of purpose (why) driving a set of attribute sub-processes (who, what, where, when) through a set of transformation sub-processes (how). All of these sub-processes also have their own sub-processes—they are a hierarchy of parts and wholes, or holons (Koestler 1967). Checkland and Scholes (1990) have proposed the CATWOE template for studying change management processes, the acronym covering the aspects of customers, actors, transformations, weltanschauung (worldview), owners and environment. They also argue that while the world is treated in hard systems as *systemic* and models of it as *systematic*, in soft systems the world is acknowledged as *chaotic* and models of it as *systemic*. The objective of soft systems models is not so much to *simulate* the world through systematic procedures, because such approaches will always be incomplete and lacking in real world richness; it is rather to *reflect on* the world in an integrated, systemic way, from the identification of problems to the implementation of change. In particular, such reflection could help to mitigate or even eliminate the unintended consequences associated with engineering projects.

Artificial Intelligence (AI) or more accurately knowledge processing can be used for or seen as providing a formalization for practice at a technical level. AI could then serve the systems approaches, in the way that mathematics has served the scientific method (Dias 2002). Both AI and mathematics are formalizations at a technical (rather than conceptual) level. Chapter 8 gives examples of how some AI techniques such as neural networks, case based reasoning, expert systems and interval probability theory can be used to capture, structure and process practitioner knowledge and experience. It also provides a philosophical grounding for practice based knowledge, drawing on two very diverse philosophers, namely Michael Polanyi and Martin Heidegger. The formalization of practice based knowledge could contribute significantly to endorsing, defending and uplifting the intellectual status of engineers.

2.6 Summary

- This study of the tensions associated with the technology versus philosophy, engineering versus science, and practice versus theory debates has helped to resolve some of the crises and answer some of the questions that engineers have in the areas of ethics, ontology and epistemology.
- In the face of the question as to whether they do more harm than good, engineers should remain proud of their contributions to society, but work at developing an acute awareness of technology's ill effects. They should also see themselves as agents of humanization as well as transformation.
- Regarding the question of whether they are managers or scientists, engineers should see themselves as holistic managers grounded in science. They should

see a whole to part relationship between engineering and science, in that engineering is richer and broader than science. The engineering role also demands a sophisticated and nuanced approach in order to deal with real world complexity.
- On the question concerning the nature of engineering knowledge, we have seen that engineers largely use practical knowledge, though they can always fall back on the theory they have been schooled in. They should however learn to see practice as being a type of theory formation too; and also work at developing some formal structures, both for the engineering approach itself and for practice based knowledge.
- The adoption of systems thinking frameworks could be useful for formalizing the engineering approach at the conceptual level. The use of knowledge processing tools such as AI may help to formalize practice based knowledge at the technical level.
- We can see that an engineer is primarily a maker (*homo faber*), a label of which to be proud, quite in contrast to Plutarch's reporting of Archimedes' views. However, strong arguments can be made as to why an engineer is very high on the scale of a thinker (*homo sapiens*) too.

Acknowledgements Adapted by permission from Springer Nature Customer Service Centre GmbH: Springer Nature: The engineer's identity crisis: *homo faber* or *homo sapiens*? by Priyan Dias, Chap. 11 (pp. 139–150) in *Philosophy and engineering: reflections on practice, principles, and process*, (Eds.) D. Goldberg, N. McCarthy, D. Michelfelder, 2013.

References

Aristotle, in *Nicomachean Ethics*, ed. by R. Crisp (Cambridge University Press, Cambridge, 2000)
D.I. Blockley, *The Nature of Structural Design and Safety* (Ellis Horwood, Chichester, 1980)
D.I. Blockley, Phil's eight maxims. Struct. Eng. **59A**(9), 292–294 (1981)
D.I. Blockley, Engineering from reflective practice. Res. Eng. Des. **4**, 13–22 (1992)
D.I. Blockley, *Bridges: The Science and Art of the World's Most Inspiring Structures* (Oxford University Press, Oxford, 2010a)
D.I. Blockley, The importance of being process. Civ. Eng. Environ. Syst. **27**(3), 189–199 (2010b)
D.I. Blockley, Analysing uncertainties: towards comparing Bayesian and interval probabilities. Mech. Syst. Signal Process. **37**(1–2), 30–42 (2013)
D.I. Blockley, P. Godfrey, *Doing It Differently: Systems for Rethinking Construction* (Thomas Telford, London, 2000)
D.I. Blockley, J.R. Henderson, Structural failures and the growth of engineering knowledge. Proc. Inst. Civ. Eng. **68**, 719–728 (1980)
P. Checkland, J. Scholes, *Soft Systems Methodology in Action* (Wiley, Chichester, 1990)
W.P.S. Dias, Reflective practice, artificial intelligence and engineering design: common trends and inter-relationships. Artif. Intell. Eng. Des., Anal. Manuf. (AIEDAM) **16**, 261–271 (2002)
W.P.S. Dias, Heidegger's relevance for engineering: questioning technology. Sci. Eng. Ethics **9**(3), 389–396 (2003)
W.P.S. Dias, Heidegger's resonance with engineering: the primacy of practice. Sci. Eng. Ethics **12**(3), 523–532 (2006)

W.P.S. Dias, Philosophical underpinning for systems thinking. Interdiscip. Sci. Rev. **33**(3), 202–213 (2008)

P. Dias, *Pompeii* by Robert Harris: an engineering reading. ICE Proc. Eng. Hist. Herit. **163**(4), 255–260 (2010)

P. Dias, The engineer's identity crisis: *homo faber* or *homo sapiens*? (Chap. 11), in *Philosophy and Engineering: Reflections on Practice, Principles, and Process*, ed. by D. Goldberg, N. McCarthy, D. Michelfelder (Springer, Dordrecht, 2013), pp. 139–150

W.P.S. Dias, D.I. Blockley, Reflective practice in engineering design. ICE Proc. Civ. Eng. **108**(4), 160–168 (1995)

J. Ellul, *The Technological Society* (Alfred A. Knopf, New York, 1948)

D. Elms, David Blockley: an appreciation. Civ. Eng. Environ. Syst. **27**(3), 175–176 (2010)

A. Feenberg, *Questioning Technology* (Routledge, London, 1999)

S.C. Florman, *The Existential Pleasures of Engineering*, 2nd edn. (St. Martin's Press, New York, 1994)

B. Glaser, A.L. Strauss, *The Discovery of Grounded Theory: Strategies for Qualitative Research* (Weidenfeld and Nicolson, London, 1967)

S.L. Goldman, Compromised exactness and the rationality of engineering (Chap. 1), in *Social Systems Engineering; The Design of Complexity*, ed. by C. Garcia-Diaz, C. Olaya (Wiley, Oxford, 2017), pp. 13–29

B. Goranzon (ed.), *Skill, Technology and Enlightenment: On Practical Philosophy* (Springer, London, 1995)

M. Heidegger, *Being and Time* (trans: J. Macquarrie, E. Robinson) (SCM Press, London, 1962)

M. Heidegger, *The Question Concerning Technology and Other Essays* (trans: W. Lovitt) (Harper and Row, New York, 1977)

A. Koestler, *The Ghost in the Machine* (Picador, London, 1967)

I.A. Monarch, S.L. Konda, S.N.Levy Konda, Y. Reich, E. Subrahmanian, C. Ulrich, Mapping sociotechnical networks in the making, in *Social Science, Technical Systems and Cooperative Work: Beyond the Great Divide*, ed. by G.C. Bowker, S.L. Star, W. Turner, L. Gasser (Lawrence Erlbaum, Mahwah, 1997), pp. 331–354

P. Nuttgens, *What Should We Teach and How Should We Teach It? Aims and Purpose of Higher Learning* (Gower Publishing Company, London, 1980)

K.R. Popper, *All Life is Problem Solving* (Routledge, London, 1999)

D.A. Schon, *The Reflective Practitioner: How Professionals Think in Action* (Temple Smith, London, 1983)

P.M. Senge, *The Fifth Discipline: The Art and Practice of the Learning Organization* (Century Business, New York, 1992)

Chapter 3
Are Failures the Pillars of Success?

3.1 All Life Is Problem Solving

Engineering is one of the oldest and most characteristic examples of a problem solving enterprise. It is concerned with transformation or useful change, and change almost always occurs when there is a felt need or problem. Sometimes of course change is brought about through seeking an opportunity for improvement, even without a perceived problem. A key feature of engineering is design, which Christopher Alexander (1964) has described as the 'elimination of misfits'—i.e. 'misfits' between a current solution and its context.

Change is frequently associated with risk. A person who takes no risks will cause little change. Part of this risk refers to the chances of not achieving what one sets out to do. Another part is the risk of unforeseen effects or unintended consequences. Thus, in the process of solving problems engineers can create new problems too. A good example is how technology is blamed for environmental degradation, both in terms of pollution and resource depletion. However, such problems can be seen as fresh 'grist for the mill'; many engineering activities are now directed towards conserving the environment. This solving of problems that gives rise to new problems and challenges is highly reminiscent of Sir Karl Popper's (1902–1994) cyclic scientific methodology. This chapter builds on the pioneering paper of Blockley and Henderson (1980) in applying Popper's ideas to engineering (see also Dias 2007).

3.2 Popper's Problem Solving Methodology

Karl Popper was one of the most dominant and influential philosophers of science in the 20th century. Scientists can perform exceptionally well without being aware of any philosophies of science (Lipton 2005); but most practising scientists would operate as Popperians, whether consciously or not. Popper (1968) considered that the main philosophical problem to be solved was that of the question concerning

P. Dias, *Philosophy for Engineering*, SpringerBriefs in Applied Sciences and Technology, https://doi.org/10.1007/978-981-15-1271-1_3

knowledge and its growth. He sought to understand the growth of knowledge in general by studying the growth of scientific knowledge, which he considered as having the fastest rate of growth. His work focused also on the scientific *method* (Magee 1973), held by him to be of crucial importance to the growth of knowledge. Other philosophers have downplayed the importance of 'method' (Feyerabend 1975), but practising scientists and engineers would largely agree that method is very important.

The popular view about method, prior to Popper's influence, was that science grew through a process of what could be called *inductive generalization*. The inductive process starts from making a succession of observations or conducting many experiments; and ends by making generalizations or arriving at theories, based on the results of those observations or experiments. It can therefore also called be an *empirical* (or practical) method. For example, we observe that a metal expands every time it is heated; also that this property is displayed by more than one metal. We can therefore make a generalized tentative hypothesis that "all metals expand when heated". We then subject this to verification in order to prove or disprove the hypothesis. Good inductivists will test a wide range of metals under different initial temperatures and ranges, and ensure that no observation violates the hypothesis, before pronouncing that the hypothesis has been 'proved'.

David Hume (1711–1776), although being an empiricist, saw that inductive generalization was not strictly logical. He pointed out that repetitions of 'particular' (or specific) observations could not logically give rise to universal theories. Any future expectations we had based on past regularities were *psychological* rather than *logical*. Theories could never be verified logically, because there would always be a chance that the sequence of regularity could cease or change. As put by Magee (1973), "from the fact that all past futures have resembled past pasts, it does not follow that all future futures will resemble future pasts".

In addition, one of the main problems with induction is its focus on *verification*. Popper (1989) argued that a more self-critical approach was required for the growth of knowledge; this would also restore a sense of humility to scientists. He proposed the following scientific methodology for replacing inductive generalization (Popper 1999; Magee 1973): (i) (old) problem; (ii) proposed tentative trial solution—a conjecture; (iii) deduction of crucial testable propositions; (iv) critical testing—attempted refutations; (v) (new) problem(s). The new problem will not arise unless and until the proposed solution or conjecture has been disputed.

The important thing in the above scheme is the recognition that science starts with a problem, and not through the 'disinterested' observation of entities. These problems could typically be gaps or anomalies in existing theories—e.g. their lack of comprehensiveness, or correspondence with observations. The above scheme is cyclic as well, in that the new problems will require new conjectures and so forth—hence, it would encourage the growth of knowledge. Furthermore, we see that the first step towards solving the problem is not observation or experimentation, but rather a proposed trial solution or hypothesis. Even if we did set out on some initial observations or experimentation, Popper said that we had, even sub-consciously, some half formed or tentative solution that we were trying out—in other words, a *conjecture*. Such conjectures could be based on our background knowledge; but imagination or aesthetics could be allowed to shape them too.

Once a conjecture or hypothesis was stated, some implications of this hypothesis could be arrived at by the process of *logical deduction*. The deductions made were not to be trivial ones, but those that were crucial for testing the hypothesis or extending its scope. Such deductions could then be tested by observation or experiment. So the movement here is from theory to experiment—the opposite of induction. The approach is also very self-critical, because Popper said that such testing should constitute attempted *refutations* of conjectures, or *falsifications* of hypotheses. We see from this that Popper is promoting falsification rather than verification. He was against verification, first because it was logically erroneous. No amount of confirming instances of tests could 'verify' a theory, because it could always fail such a test in the future; on the other hand, even a single failed test could falsify it. There was thus an 'asymmetry' between verification and falsification (Popper 1983). Also, a focus on verification would not promote growth in knowledge, whereas the replacement of a theory by a better one would.

Let us move from the *overthrowing* of theories to another way in which knowledge can grow. This is by the *improvement* of theories. Such improvement too could be assisted by a falsificationist outlook. Magee (1973) gives a good example of this. Suppose we state the hypothesis that "water always boils at 100 °C". If we had a verificationist outlook, we would merely be content to find many confirming instances of water boiling at 100 °C. However, a falsificationist would try to devise critical and novel methods of trying to falsify this hypothesis. Such a person would probably discover that water does not boil at 100 °C at elevated altitudes or in closed vessels. The hypothesis would then need to be framed in a more comprehensive and fundamental manner, and perhaps give rise to the idea that "water boils when its saturated vapour pressure is equal to that of the surrounding atmosphere"—thus resulting in a growth of knowledge.

It should be noted that Popper (1968) also used the concept of '*falsifiability*' as a criterion of demarcation to judge whether a body of knowledge could be called 'scientific' or not. In other words, for a discipline to call itself a 'science', its theories had to be enunciated in a way that it was possible to test and falsify them. On this basis, he refused to admit that the social sciences (in particular Marxism) and psychoanalysis were sciences, because their proponents tended to 'explain away' disconfirming instances; or expand their theories arbitrarily to accommodate such disconfirmations, rather than changing them.

3.3 Extending the Methodology

Solving a problem by first positing a conjecture was a scheme that Popper said applied to all knowledge. Popper broadened his theory of scientific knowledge to encompass our entire evolutionary history, which to him was the growth of knowledge through problem solving. He drew parallels between the 'random' genotype variations in

Darwinian evolution and the 'bold' conjectures in his scientific methodology, for which an element of 'irrationality' was virtually indispensable (Popper 1968); also between the 'survival of the fittest' at the phenotype (or organism) level as a result of a harsh environment and the critical testing of propositions for refuting theories in his methodology. He also called this process 'error elimination' (for both species and knowledge evolution). Popper compared the now discredited Lamarckian theory of evolution (where changes in species were supposed to accumulate at the phenotype level) to the idea of knowledge accumulation through induction, which he strongly disputed. The other feature of importance, whether of biological or scientific evolution, was the crucial ingredient of feedback (Popper 1972). A cyclic methodology such as Popper's strongly emphasized learning from feedback. It was such feedback that resulted in error elimination and hence progress. Popper considered that the growth of knowledge formed a seamless web "from the amoeba to Einstein", the amoeba being eliminated by evolution when it *makes* mistakes so that more robust species can evolve; whereas Einstein *looks* for mistakes in order to improve or even discard his theories in favour of better ones (Popper 1999).

The idea of feedback leads us to another area of Popper's inquiry we shall look at—that of social change. Popper's ideas of critical rationalism, error elimination and progress of knowledge in the evolutionary and scientific domains led to his arguing for an 'open society' in the political sphere. He was particularly opposed to shaping society based on predictions. "If it is possible for astronomy to predict eclipses, why should it not be possible for sociology to predict revolutions?" he asked (Popper 1960; Dias 2014); and proceeded to show why not. For one thing, he did not consider theories of society to be as dependable or testable as theories concerning science, since the former involved much greater complexity, including the essential unpredictability of human beings. It could be argued that such historical 'laws' were merely generalizations of trends and not really laws (Popper 1960). There are similarities between such 'trends' and the generalizations made in scientific induction to arrive at universal theories, a method rejected by Popper as we have seen. Induction can also be compared to generating a relationship between a dependent and independent variable by curve fitting after a series of observations. The best we can do through such fitting is to discover a *trend*, and not a *law* based on some genuine natural phenomenon, such as energy for example.

Popper was therefore opposed to shaping the long term future of society based on some grand utopian scheme, as espoused by Plato or Marx, both of whom he labelled as 'enemies of the open society'. Popper argued that change had to be 'engineered' from a given situation, calling it 'piecemeal social engineering' (Popper 1960). This has similarities with starting from a problem situation in Popper's scientific methodology, and with the idea that the growth of knowledge is evolutionary, with errors being eliminated through 'feedback' during each cycle of growth. How then could we ensure 'error elimination' in society? For one thing, Popper advocated that it be possible for those who rule to be removed periodically. He said therefore that the important question was not "Who should rule?" as raised and pursued by Plato and Marx, but rather the question of "How can rulers be removed?" (Popper 1999). For another, Popper valued a society with diversity. He considered the lack of a unifying

Table 3.1 Common themes from Popper across Evolution, Science and Society (from Dias 2007)

Domain →	Evolution	Science	Society
Problem domain (P)	Survival	Knowledge	Governance
Creativity of trial solution (TS)	Random variation in genotype	'Irrational' elements in hypothesis; also competing hypotheses	Pluralism in society
Error elimination (EE) method	Harsh environment (nature)	Critical testing (scientific community)	Public opinion (population)
Error elimination (EE) result	Extinction	Refutation	Change of rulers
Problem solving scheme	Natural selection (Darwinian)	Cyclic methodology: P_i → TS → EE → P_{i+1}	'Piecemeal social engineering'
Rejected alternative	Lamarckianism	Induction	Historicism
Result of alternative	–	Authoritarianism	Tyranny

idea in western democracies to be a strength rather than a weakness (Corvi 1997). Compare this to the diversity required in the gene pool for successful Darwinian evolution.

Table 3.1 presents Popper's remarkably unified ideas regarding problem solving in three distinct domains, namely evolution, science and society. All of them have a problem generating environment, and methodologies for proposing creative solutions and error elimination. The ingredients for incorporating creativity in solutions were considered by Popper to be very important; it was these elements that rescued his methodology from the bonds of induction. The critical approach to error elimination was equally or more important. The pluralism, public opinion and change of rulers required in the social sphere led Popper to argue for liberal democratic political systems. The problem solving scheme for the scientific domain consisted of his cyclic methodology of old problem → trial solution → error elimination → new problem. This had parallels in the other two domains too. Although his methodology was conceived for replacing the alternative of induction in science, similarly unsatisfactory parallels for induction were identified by Popper in the other two domains. Notturno (2000) has argued that an inductivist outlook could lead to authoritarianism in scientific institutions, while Popper himself was very clear about the link between sociological generalizations and social control, which led to tyranny.

3.4 Cyclic Engineering Processes

It is not only the *idea* of problem solving that makes Popper relevant to engineers; his *mode* of problem solving—i.e. a *cyclic* mode—is also something that is frequently used by engineers, especially in design. Engineers in design organizations would like to make design a linear process, i.e. to start with a set of specifications and end up with a solution. But in principle and often in practice, design is a circular process. Figure 3.1 presents the design cycle of specifications—synthesis—analysis—evaluation and compares them with the key elements of Popper's methodology. We shall consider each element in turn.

We start with the *specifications*, analogous to Popper's 'problem'. An engineering problem has to be formulated or defined through a set of specifications that have to be

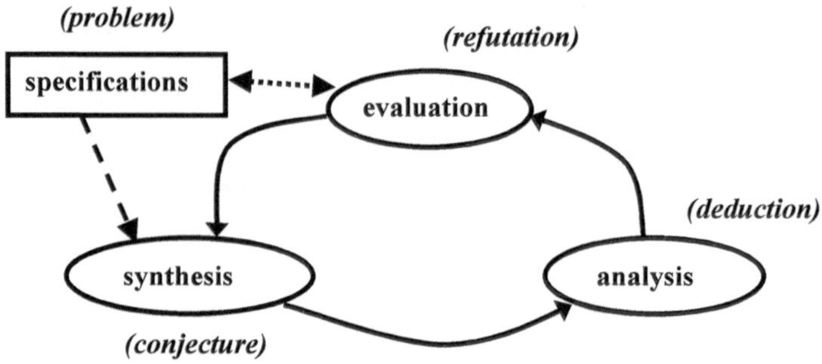

Fig. 3.1 Engineering design methodology and Popper's problem solving methodology (from Dias 2007)

satisfied by the design. Specifications can be seen as performance criteria or functions expressed in engineering terms. Conflicting specifications are often encountered in engineering; so the problem is not merely to satisfy the specifications, but also their conflicts. For example, an architect may want a beam not to exceed a specified depth so that the ceiling provided under the beam will have a sufficient clear height above the floor. On the other hand, the services engineers may want the air conditioning duct also placed within the ceiling. This is quite apart from the engineering specifications that the beam itself has to satisfy—i.e. that it should be strong enough to carry the loads imposed on it without deforming excessively.

The next step in design is to posit a solution, similar to a Popperian conjecture. This is called *synthesis*. The step from specifications to synthesis is often a difficult one. This is where creativity in design comes into play, comparable to Popper's 'irrational element' that is required for a conjecture—for example proposing a beam with a pierced web in order to satisfy the architectural, services and engineering specifications described above. Factors such as memory (of *past* designs), perception (of the *present* context) and scenarios (of *future* usage or deterioration) are all required for synthesis (Blockley 1992). Synthesis is difficult to automate and in principle cannot be done via an algorithm or mathematical model. There is also the issue that design involves abductive reasoning or finding an 'inverse solution' (see Sect. 2.3)—where there may be more than one way to meet a single set of specifications. The managing of conflicting specifications via creative solutions (e.g. a pierced beam) is also part of synthesis.

If the design is a fairly routine one, experienced engineers will be able to make a guess at a good initial solution which may not even need to be changed subsequently. Younger engineers may be able to use 'rules of thumb' (frequently referred to in Chap. 2) that have been developed over the years. Nowadays they may be able to resort to Artificial Intelligence techniques (see Chap. 8); these can help us to propose solutions based on past experience (Dias 2002; Dias and Padukka 2005). Creative solutions can sometimes be generated using techniques of Artificial Life, an example of which is described at the end of this chapter—these techniques are often based on a Darwinian metaphor, giving us another link to Popper.

Once we have synthesized a solution, we can resort to the *analysis* of that solution. Analysis is well supported by mathematics and software. This constitutes, to a large extent, the engineering science component of the engineering process (see Fig. 2.2). For civil, structural and mechanical engineering it is underpinned essentially by Newtonian mechanics. The process of analysis can be compared to Popper's testable deductions, which are derived from the conjectured hypothesis.

The next step is to compare the results of our analysis with the specifications—a process of *evaluation*. If our original solution was either unsafe or too conservative, this comparison will show it up. This is the engineering parallel to Popper's idea of refutation. Engineers may think they are checking or trying to verify that the solution they proposed is satisfactory. But especially when trying to optimize their designs, a better way to think of evaluation is that it is a way to refute their proposed syntheses; or 'falsify' them, in Popperian language. Consider the design process for optimizing the form of the Tianjin CTF Finance Centre for wind loading using a wind tunnel (Cammelli 2018); a total of 17 different aerodynamic solutions were developed and tested in an interactive fashion. Another way to design is by comparing solutions that are proposed as candidates—for example the four alternatives tested for the Ridgeway Footbridge (Willoughby 1996). Both in optimizing and comparing alternatives we see that proposed solutions are in fact being rejected after critical testing. Trial and error procedures are very characteristic of engineering (Vincenti 1990), even during actual construction. For example, geotechnical design solutions are sometimes arrived at via feedback from monitoring the unfolding soil response (LeMasurier et al. 2006).

In Fig. 3.1, the comparison with the specifications is shown with a double-headed dotted line. Normally, the specifications are held as the standard against which a design has to be measured, and failure to comply will demand a fresh synthesis (i.e. a new conjecture). However, if the specifications have been defined too tightly or with conflicting constraints, they may need to be changed or relaxed. Clearly this has to be done with the consent of the client, while taking into account the safety of the public.

It should be noted that the specifications are not really part of the main cycle in Fig. 3.1, especially in the more routine design situations where specifications are not changed after evaluation. The core engineering design cycle is synthesis—analysis—evaluation. Some authors (Coyne et al. 1990) denote this cycle as analysis—synthesis—evaluation. The analysis referred to by them is problem analysis, inclusive of defining specifications. It requires broad engineering knowledge and an appreciation of the 'big picture'. In Fig. 3.1 this type of analysis is subsumed under 'synthesis' (and perhaps specifications). The 'analysis' referred to in Fig. 3.1 on the other hand is the narrower type of analysis that can be tackled by engineering science techniques (e.g. mathematical models); but it also includes the creative envisaging of future scenarios for which the synthesis has to be analyzed.

It may also be noticed that the synthesis, analysis and evaluation cycle refer mostly to activities rather than states, with the arrows between them in Fig. 3.1 merely serving as links between activities. The design cycle can be depicted using states too; then it is the arrows linking them that will denote activities or transformations. In order to do this, we shall build on the well-known triad of 'function—structure—behaviour'; and use the following symbols, after Umeda et al. (1990) and Hybs and Gero (1992):

F: function
S: structure (i.e. synthesized design object)
P: product (i.e. fabricated object)
Be: behaviour that is expected (based on specifications)
Bs: behaviour of the design object in an appropriate (mathematical) model
Ba: behaviour that is actually exhibited by the fabricated product

The following transformations can be identified:

F → Be: specification
Be → S: synthesis
S → Bs: analysis
Bs ↔ Be: evaluation
S → P: fabrication
P → Ba: operation
Ba ↔ Be: diagnosis

The transformations involved in the activities of specification, synthesis, analysis and evaluation are made explicit here. In addition, the wider cycle of the *world* in which the design object is realized, becomes evident (see Fig. 3.2). The role played by the behaviours is of crucial importance. They are the means by which comparison and correction are carried out—note the double headed arrows above. We saw earlier that comparison within the design cycle was called evaluation; this leads to improved design objects through repeated syntheses. On the other hand, comparison in what we shall call the *world cycle* can be called *diagnosis*; this should lead to improvements in the entire design cycle, or the 'calculation procedure model' (Sect. 2.3), as it has been called by Blockley and Henderson (1980). Diagnosis could indicate shortcomings in fabrication or operation too, but the sensitive designer should try to design against as many fabrication and operational errors as possible with an 'idiot-proof' design (Dias 1994).

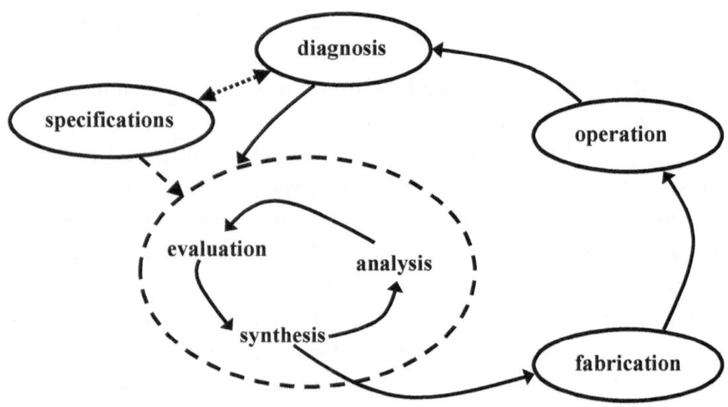

Fig. 3.2 Design cycle embedded within world cycle (from Dias 2007)

We are now in a position to compare the two cycles, i.e. the design cycle and the world cycle, with Popper's cyclic methodology for the growth of scientific knowledge—see Table 3.2. The problems (both old and new) for the two cycles differ, being the specifications for the design cycle and a failure in the 'calculation procedure model' (CPM) for the world cycle. The crucial difference is in the conjecture. In the design cycle, it is the synthesized design object that is the conjecture (analogous to a scientific theory), which is subjected to refutation through evaluation. In the world cycle, it is the entire design cycle or CPM (which is a closer analogy to a scientific theory) that is being tested (by diagnosis) with respect to its performance in the world (inclusive of construction and operation in that world). The world cycle is therefore a much wider and longer term one—since failures may not show up for a long time—and embeds within it the design cycle (Fig. 3.2). The knowledge generated through the world cycle is long term and industry-wide in nature; whereas in the design cycle it is project-specific and short term—i.e. in the process of trying to test our design object against various failure modes through a CPM. In the design cycle, we generally do not question the validity of the CPM. It is the *design object* that is being tested for fitness. In the world cycle on the other hand, *the entire CPM*, inclusive of all mathematical and heuristic models used, is being judged.

Learning through feedback in cyclic processes is very much a part of management practice today (Senge 1992; Checkland and Scholes 1990), for example as embodied in the art of reflective practice (Blockley 1992; Schon 1983; Dias and Blockley 1995). Recall that engineers are more managers than scientists (Sect. 2.3). Consider the representation of an organizational or even industry-wide activity. Although it can be seen as a linear process of converting inputs to outputs and waste, as in Fig. 3.3a, it is probably more realistic and helpful to view it as a circular process, as in Fig. 3.3b. The latter representation portrays the idea that the organization does not remain static during its operations—it changes too; and hierarchically nested reflection can lead to improvements, at one level through learning by reflecting on the transformation process, and at a deeper level by reflecting on and evaluating the

Table 3.2 Popper's cyclic methodology compared with the design cycle and world cycle (from Dias 2007)

		Design cycle	World cycle
Popper's categories	(Old) problem	Specifications	Failure
	Conjecture/hypothesis	Synthesis or design object	Calculation procedure model or design cycle
	Testable deductions	Analysis	Fabrication and operation
	Refutation/testing	Evaluation	Diagnosis
	(New) problem	Specifications	Failure
Nature of growth in knowledge	Extent	Project-specific	Industry-wide
	Duration	Short-term	Long-term

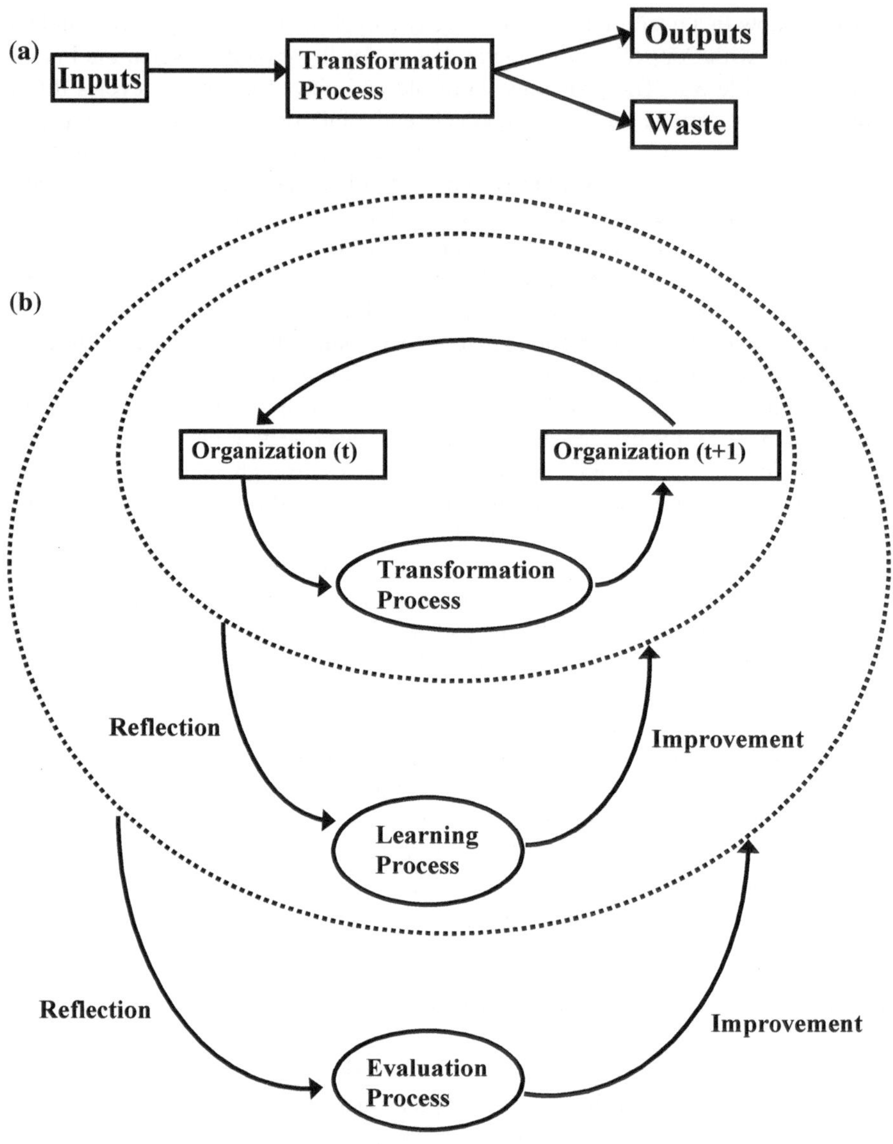

Fig. 3.3 Organizational process viewed as **a** linear and **b** circular (from Dias 2007)

learning process itself. Learning within organizations can be promoted by asking the question "*What* did we learn from our experience?" from time to time; while evaluation can be a more structured exercise carried out periodically (e.g. annually) by asking the question "*How* did we learn during our experiences?" or "*How* can we improve our learning process?".

3.5 The Role of Failure in Engineering

Popper's great innovation was that a theory could only be falsified, as opposed to being verified. If we consider the design cycle, where the synthesized design object corresponds to Popper's conjecture, we have seen that falsification is precisely what we actively try to do. Engineering is steeped in a culture of safety. This means that we are always trying to identify modes of failure, in order to design against them. Henry Petroski's book *To Engineer is Human*, is subtitled *The role of failure in successful design* (Petroski 1969); one of its chapters is called 'Success is foreseeing failure'. For the structural engineer, the 'ultimate' failure is collapse. However, a structure, or any other engineering product, can fail in many other ways.

The ultimate limit state (i.e. the point beyond which failure would occur) is certainly the most *important* one in structural engineering. However, a serviceability limit state—i.e. a point beyond which a structure would not be useable (e.g. by exhibiting excessive cracking or deflection)—could be more *critical*, in that it could be the criterion that governs the design. This is very common in prestressed concrete structures and water retaining concrete structures, where the amount of steel wires or bars required for the limiting of crack widths to below a desired value is often greater than the steel required for preventing collapse. What happens in practice is that engineers make the initial synthesis based on their background knowledge regarding what the critical limit state is. If the design checking satisfies the criteria for that limit state, the chances are that all other limit states, when subsequently checked, would be satisfied by that solution as well. The difficulty is in obtaining a good initial solution, so that it will not need to be changed many times. Design can indeed be seen as cyclic, and the fact that many analyses and design checks need to be done reflects that cyclic nature. However, designers would prefer not to make changes in the synthesis or design object after each test. The choice of good initial solutions depends on accumulated engineering knowledge and reflection thereon (Dias and Blockley 1995); this can be based on an engineer's own experience or that of his company or even that of the entire industry. Later on in the book (Chap. 8) we explore how Artificial Intelligence techniques can capture and process such experience.

An important aspect of foreseeing failure is that we try to design not only for the 'normal', but also for the 'abnormal'. There are two ways in which we can deal with unforeseen disasters. The first is to build redundancy into the system. In structural engineering this is equivalent to providing alternative load paths—i.e. ensuring that there can be sharing of the load that is shed if one element fails accidentally. The other is to make sure that failure takes place safely; in other words, to provide *fail-safe* mechanisms (Dias 1994). In structural engineering, this is achieved through ductile failure modes. We design structures so that if there is unanticipated overload, the elements will exhibit cracking and deflection prior to collapse. This will provide warning for inhabitants to take remedial action or vacate the building. Ductility is also required for redistributing any loads shed by failed members to alternative load paths.

We could even think of ‘ductility’ at the structural level, by trying to arrange for members ‘higher up’ in the load chain to fail first. In addition to providing warning as before, this will prevent the catastrophic failure of more heavily loaded members ‘lower down’ the chain by reducing the load on them—i.e. the load previously carried by the failed member (Dias 1994). For example, there is a documented failure of a 32 m span prestressed concrete girder at a cement works due to dust accumulation on the roof (Dias et al. 1994). The load from the dust accumulation was transferred to the girder first via roofing sheets that spanned just 1.2 m between concrete purlins, that in turn spanned 9 m between the girders. If the sheets or the purlins had failed first, the large scale failure could have been avoided. The concept here is like the electrical engineer’s fuse in an electrical circuit; in order to prevent the failure—i.e. melting of the wires in the circuit due to an unforeseen electrical overload—a pre-designated element, i.e. the fuse, will melt first because its thermal capacity is made to be lower than that of the wires. If such failure occurs, we merely replace the small fuse rather than having to re-wire the entire circuit; the ‘failure’ is much smaller and easily managed. Micro circuit breakers have now replaced fuses, making failure even easier to deal with.

So we see that within the design cycle, the aim is to actively falsify our conjecture (in the form of a design synthesis), in order to improve on it—very much a Popperian objective. One of the ways in which such falsification takes place in science is if there are *alternatives*. This is true in engineering design too—especially of large or novel products or systems. More than one solution to the specifications will certainly be generated at least at the conceptual design stage in the design cycle. The alternative that does better in the design checks (tests) is the one adopted for fabrication in the world cycle—see also Sect. 3.4.

How about the world cycle? Can we seek active falsification there too? Recall that our conjecture here is the design cycle or the calculation procedure model (CPM)—see Table 3.2. It would be nice to find ways to test the accuracy of the CPM; in other words, can we build a structure and subject it to failure, to check how accurate our CPM is? Automobile manufacturers try to do this by their ‘crash tests’. This is because an automobile is a mass produced artefact and testing a few to failure can be tolerated in order to fine-tune the associated CPM—although here too the actual performance during or after prolonged use cannot be tested. Buildings and bridges are however one-off artefacts; each one is unique, at least with respect to the ground it is founded on. Especially in structural engineering therefore, we cannot actively seek failure in the world cycle, because engineering is carried out within a social matrix. If such engineering products are fabricated to test the ‘truth’ of the CPM by testing them to ‘failure’, the public would be greatly inconvenienced, if not incapacitated or even killed. Technology underpins the entire fabric of our society. We have to make sure, as far as possible, that there are no breakdowns.

In other words, we are not interested primarily in the *truth* of our CPM, but rather about whether it constitutes a *safe* and *dependable* method for fabricating products that will be operated in the world (Blockley 1980; Blockley and Henderson 1980). Science on the other hand is carried out within the confines of a laboratory, where theories can be put to the test and active falsification sought. This is why scientific knowledge grows so rapidly. These distinctions between scientific theories and engineering models are further described in the next chapter (Sect. 4.5). In ancient days, there were strict penalties that were imposed on builders who did not deliver safety, as exemplified in the code of Hammurabi, 6th king of Babylon (1792-50 BC) and cited in Petroski (1969):

> If a builder build a house for a man and do not make its construction firm and the house which he has built collapse and cause the death of the owner of the house, that builder shall be put to death.
>
> If it cause the death of the son of the owner of the house, they shall put to death a son of that builder.
>
> If it cause the death of the slave of the owner of the house, he shall give to the owner of the house a slave of equal value.
>
> If it destroy property, he shall restore whatever it destroyed, and because he did not make the house which he built firm and it collapsed, he shall rebuild the house which collapsed at his own expense.
>
> If a builder build a house for a man and does not make its construction meet the requirements and a wall fall in, that builder shall strengthen the wall at his own expense.

However, if and when engineering failures do occur, they provide invaluable opportunities for improving or even overthrowing our CPM, and thus for industry-wide growth in engineering knowledge. Engineers are therefore very interested in failures, since they provide scope for learning. It has been argued (see Fig. 2.2) that the CPM could be seen as having an engineering science core, surrounded by 'shells' of idealization, margins of safety, design philosophy, design context and engineering process. While the engineering science core is now mature and unlikely to grow very much, there is still much scope for learning in other aspects, as we shall see in the next section, which gives an example each of failure in the six components of the CPM. It must be appreciated that reflection on over-conservatism and unnecessary use of materials—i.e. the opposite of failure—also leads to changes in the CPM. The availability of powerful structural analysis programs for example, reduces to a large extent the need to make conservative idealizations. Also, the use of reliability theory has been used to reduce margins of safety, where they have been unnecessarily large (Beeby 1994).

3.6 Failures in Various Components of the CPM

We start with a defect in an *engineering science* theory described by Petroski (1992) and discussed by Dias and Blockley (1994). It has to do with Galileo's wrong assumption that a cantilever behaved like a beam hinged at the bottom edge of its fixed end, with a uniform tensile stress σ across its section resisting the overturning moment caused by the load W at the end of the cantilever (see Fig. 3.4a). The maximum load that could be carried can then be obtained as $W = \sigma bh^2/2L$, where h, b and L are dimensions of the cantilever as in Fig. 3.4a. The correct stress distribution however

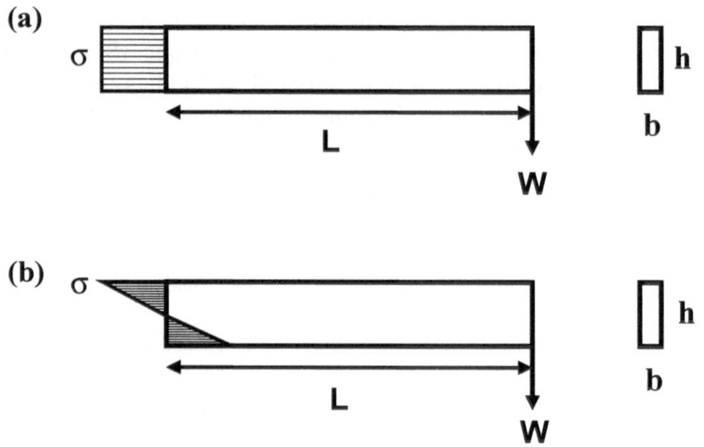

Fig. 3.4 Maximum longitudinal stress, σ, in cantilever beam: **a** as per Galileo; **b** actual (after Petroski 1992)

is shown in Fig. 3.4b, where σ is the maximum stress and the corresponding value of maximum load is $W = \sigma bh^2/6L$, which is three times lower than Galileo's estimate.

Even though Galileo's theory would have been used in real structures such as timber ships for example, given that Renaissance shipbuilders were likely to have used factors of safety in excess of 3, Galileo's error was not discovered. This shows that the safety of a structure does not guarantee the truth of the CPM. Blockley (1980) says that the CPM, as realized in most structures, is only 'weakly not falsified'. There may be components in the CPM, such as Galileo's hypothesis in this case, that are erroneous, even though structures based on it do not collapse. The great pity where Galileo's hypothesis is concerned however, is that it could have been falsified and corrected by careful and critical laboratory testing; for example by testing the beam in axial tension, in order to ascertain the value of σ. This example shows the importance of a critical and falsificationist approach towards our hypotheses, as advocated by Popper.

Figure 3.5 shows the schematic view of a roof truss with two end supports and a central one. In normal practice, such trusses are constructed with a (laterally) free or sliding end. This is to allow the truss to expand and contract with variations in temperature without setting up any internal stresses. Also, the assumption of fixed conditions at both ends would require full lateral restraint at end supports—a condition difficult to achieve in practice. The truss however, had been idealized during analysis as having both ends fixed, because it was to be rigidly welded to the tops of steel columns. The result of this is to change the structural action from the usual continuous beam to an arch; and the stress state in the bottom chord from the usual large tension at the two midspans (with large compression near the central support), to a small compression throughout. Therefore, the truss had been fabricated with a fairly small bottom chord section. However, the relatively slender columns to

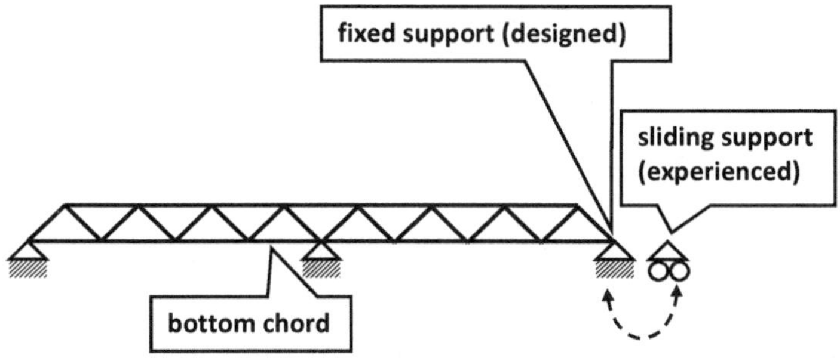

Fig. 3.5 Idealization of roof truss (after Anwar 1997)

which the truss was welded would have deflected laterally due to the truss loads; and the trusses actually behaved as continuous beams (and not like arches as idealized). This caused the bottom chord to fail either in compressive buckling near the central support or in tension at the site fabricated splices near the midspans (Anwar 1997).

This was clearly an error in *idealization*. Changes in boundary conditions can cause large differences in internal stresses, and such mistakes can very easily be made in today's design climate, where many easy-to-use computer software packages are available for structural analysis. This is a cautionary tale, not only with respect to the need for correct idealization, but also with respect to the proper use of software (MacLeod 1995). Idealization errors can occur for loads too—for example, idealizing snow loads as uniformly distributed ones may not capture the more onerous case of a varying load, and can cause collapse (Pidgeon et al. 1986).

A good example of a failure due to an inadequate *margin of safety* is the collapse of the Dee Railway Bridge in 1846, reported by Petroski (1994) and Sibly and Walker (1977). The Dee Bridge was made of cast iron girders. These were sized at the time based on a formula by Hodgkinson, which gave the central load that could be carried by such a girder of given length, depth, and area of tension flange. The emphasis was on the tension flange because cast iron was much weaker in tension than in compression; and cast iron girders were fabricated with a large tension flange but a small one in compression. By the time Robert Stephenson designed the Dee Bridge, around 15 years of construction using such girder bridges had elapsed; and spans had increased from around 25 feet to the 98 foot spans of the Dee Bridge. Although it was common practice to use global factors of safety of around 3–4, Stephenson chose to use one of only 1.5, since he thought that he was also strengthening the girders with wrought iron trussing (see Fig. 3.6). One of the three bridge spans collapsed when a train was crossing it.

In the subsequent inquiry, contemporary engineers suggested factors of safety ranging from 3 to 7. The problem with Stephenson's factor of safety was not only that it was rather insufficient in itself, but that it was also clearly inadequate in the context of the level of understanding regarding the interaction between the wrought iron truss and the cast iron girders. It is likely that the prestressing effect of the wrought iron ties created additional bending and compression in the girders, causing

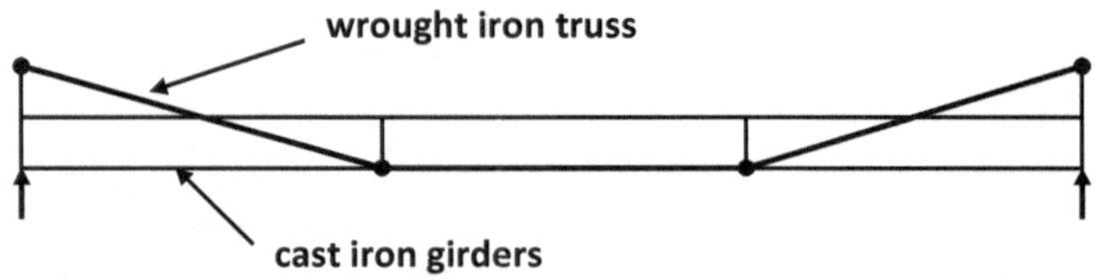

Fig. 3.6 Main structural elements in Dee Bridge (after Petroski 1994)

them to fail by lateral torsional buckling, given the small compression flange. This problem would have been exacerbated by the geometric imperfections in terms of out-of-straightness in Stephenson's long girders. Such imperfections and stress levels in shorter girders would not have been sufficient to cause failure by lateral torsional buckling. However, this second order effect became the critical mode of failure for the long spans of the Dee Bridge, especially with the reduced factor of safety. An undamaged girder from the Dee Bridge tested by Stephenson was found to fail at a load much lower than that predicted by the Hodgkinson formula. Given that the Dee Bridge was the longest of its kind to be constructed, it could be argued that a higher margin of safety should have been used, especially because the truss-girder interaction was not well understood. Hence we could say that the margins of safety component was falsified by this catastrophe.

One of the most significant structural failures of the 20th century was the partial collapse in 1968 of the 22 storey Ronan Point apartment block in Camden Town, East London in the U.K. (Levy and Salvadori 1992). The block had been constructed in the post war years using precast elements, which enabled the city to be reconstructed quickly after the war bombing. Both loadbearing walls and floors were made of precast concrete panels that were factory-cast and assembled on site. Early one morning at around 5:45 am, a resident on the 18th floor had attempted to light a gas stove; whereupon there was an explosion as a result of an unnoticed gas leak. The explosion pushed out some loadbearing walls on that floor. This caused the walls above that floor, and the slabs they supported, to collapse, since they had now lost their bearings. The resulting debris on the 18th floor kitchen slab caused that slab to collapse as a result of overloading, and precipitated similar collapses on all the lower floors. Four persons were killed, a number that could have been much greater had the explosion been at a later time of day. This brought into focus two issues with respect to *design philosophy*—one, the importance of structural integrity, and the other, the principle that the consequences of an accident should not be disproportionate to the cause. As expressed lucidly in a previous British code of practice (CP 110: 1972):

> The layout of the structure in plan, and the interaction between the structural members, should be such as to ensure a robust and stable design: the structure should be designed to support loads caused by normal function, but there should be a reasonable probability that it will not collapse catastrophically under the effect of misuse or accident. No structure can be expected to be resistant to the excessive loads or forces that could arise due to an extreme cause, but it should not be damaged to an extent disproportionate to the original cause.

As a result, designers today are very conscious of the above ideas. For example, in the subsequent British code of practice for concrete (BS 8110: 1997), all provisions for ensuring stability and robustness were highlighted in a flowchart. The main approach to ensuring robustness and preventing progressive collapse is to ensure that all concrete elements are properly tied by reinforcement. If they cannot be so tied, the designer has to assume that each untied element is 'lost' in turn, and ensure that the rest of the structure can still stand up, albeit under reduced loads and/or factors of safety. In fact, for structures designed against bomb explosions, elements are generally considered 'lost' in turn, whether or not they are tied to the rest of the structure.

We next consider a failure in design *context.* A part type plan for two storey school buildings in Sri Lanka is shown in Fig. 3.7. During the 2004 Indian Ocean tsunami, the oncoming and receding waves caused severe scoring in the corner column foundations in two such structures, causing collapse of the end bays. These buildings with two lines of parallel column and pad footings had thitherto performed well under a range of conditions. However, it failed in the context of some locations on the eastern Sri Lankan coast that directly faced the tsunami-generating Sumatran fault. Figure 3.7 shows how the introduction of a new column and infill walls in end bays could make the design more robust in such contexts (Dias et al. 2006).

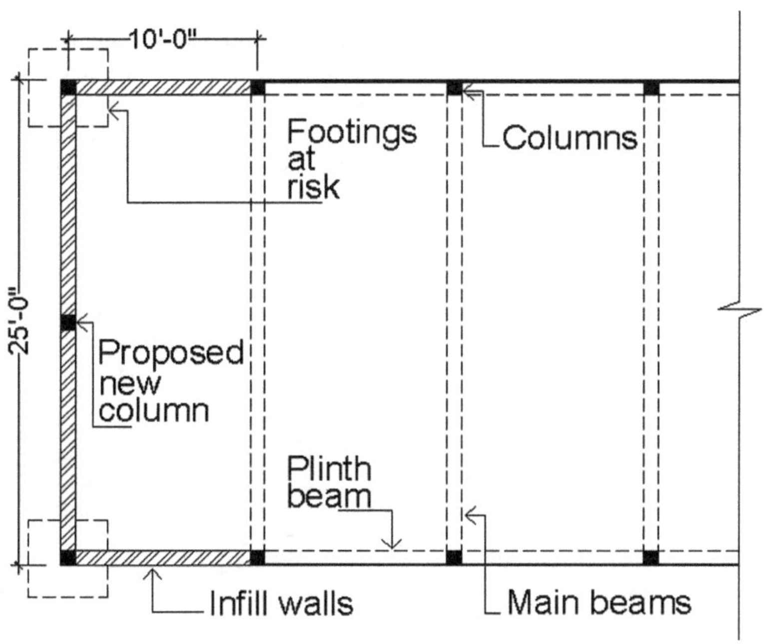

Fig. 3.7 Proposed new column and infill walls in type plan of school building to prevent failure in conditions where scouring is expected (after Dias et al. 2006)

Perhaps the most complex component of the CPM is the *engineering process*. This involves, among other things, unambiguous communication, especially of design intent. The failure that best illustrates this is the collapse of the suspended walkways at the Hyatt Regency Hotel in Kansas City, Missouri, U.S.A. in July 1981, shortly after it was constructed (Levy and Salvadori 1992). The 120 foot long walkways were at 2nd, 3rd and 4th floor levels, and enabled residents in the main residence block (with guest bedrooms) to access a function block (that housed meeting and dining rooms) without having to go down to the atrium. The walkways were suspended by hanger rods from the roof of the atrium. The 3rd floor walkway was on one side of the atrium; while the 2nd and 4th floor walkways were on the opposite side, with one hung under the other. During a well-attended dance competition in the atrium, which guests watched from the walkways as well, the 2nd and 4th floor walkways collapsed together into the crowded atrium, killing 144 persons and injuring over 200 more.

Upon investigation, it was found that the main cause of collapse was the fact that the walkways had not been hung as originally intended by the designer. The main structural elements of the walkways were two wide flanged beams that ran on both sides of a walkway along its length. These were supported, at 30 foot intervals, by transverse box beams, obtained by welding together two channel sections. Vertical holes were drilled through the ends of these box beams to pass the hanger rods. In the original design, the same hanger rod passed through both 2nd and 4th floor walkways, with nuts and washers below the beams holding the walkways to the hangers (see Fig. 3.8a). In the drawings submitted by the contractors however, this detail had been changed to that shown in Fig. 3.8b, where the lower 2nd floor walkway was carried not directly on a 'through' rod, but via a rod segment by the 4th floor walkway. The drawings had been stamped 'Approved' by the architects, and 'Reviewed' by the structural engineers. It transpired later that the change had never been checked,

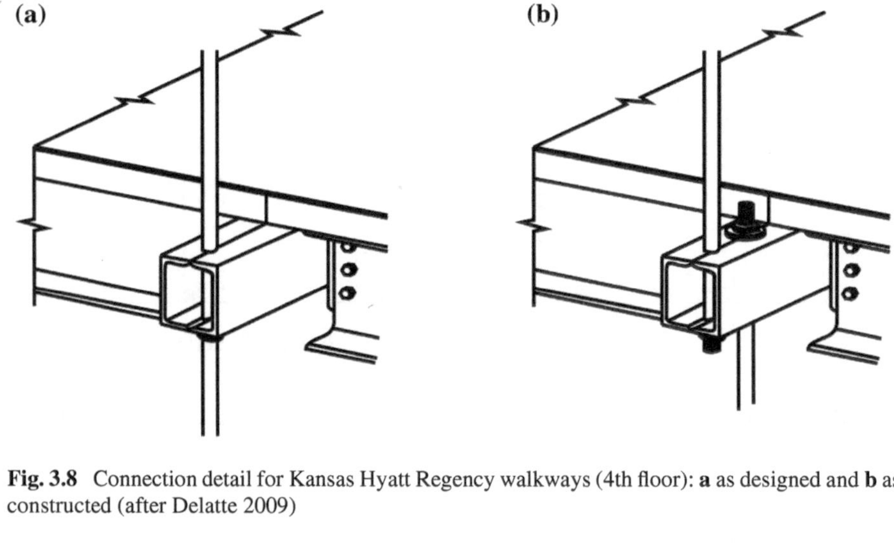

Fig. 3.8 Connection detail for Kansas Hyatt Regency walkways (4th floor): **a** as designed and **b** as constructed (after Delatte 2009)

presumably because it was considered a minor 'detail'. As it happened, this change in detail completely changed the load carrying behavior of the walkway system.

The original design was analogous to two men hanging one under another on the same rope. The change corresponded to the lower man hanging on the upper man. The grip of the upper man on the rope now becomes all important, and would be the weakest link. The strength of the rope is not the critical issue. In the same way, failure was initiated in the walkway system (as discovered in subsequent testing), by the nut at the bottom of the 4th floor walkway box beam punching through its hole. It was also discovered that there was no redundancy built into the design. Once a connection was broken, other hanger rods did not have the capacity to carry the extra load, despite the fact that the number of people on the walkways was well below the design loading. In addition, it was found that the capacity of the nut and washer connection was only 60% of the prescribed value.

This failure teaches us many lessons, for example that more than one error was involved in the failure; and also the importance of details. However, the main lesson it teaches us is the importance of proper communication. The contractor did not understand the significance of hanging both walkways on the rods independently. And most outrageously, the designer did not see the implications of the change in detail made by the contractor. Also, the contractor may have made the change because constructing to the original detail could have been very difficult. It would have involved very long hanger rods, and may have necessitated cutting threads in those rods all the way from the 2nd floor to 4th floor level, quite apart from threading nuts through that length. This shows that perhaps the designer was not designing for constructability. All of this is part of the design cycle. So, in this failure, above all it was the engineering process component of the CPM that was falsified. There is much greater emphasis now on improving the design process and promoting teamwork among different players in that process.

3.7 The Genetic Algorithm for Optimization and Design

Popper's evolutionary scheme involved not only a cyclic process, but also an element of *randomness*, especially when he brought Darwinian evolution also into his general theme of 'error elimination'. Here too, there are parallels in engineering, particularly through its embodiment in the genetic algorithm. The use of genetic algorithms is not confined to engineering although its pioneers, Holland (1975) and Goldberg (1989) were both engineers. There are two broad areas in which the genetic algorithm has been used, the first for optimization and the second for generating novel solutions in design.

Both optimization and design can be considered as *search* problems. In both cases there is the problem of having a large number of parameters, each of which could have a large number of values or states. In optimization, the problem is how to search this vast multi-dimensional space so that an optimal or near optimal solution is obtained. In design, the problem is how to select combinations of parameters that will constitute novel or creative designs. It should be noted that cycling through the design and world cycles, as described before, will probably result only in incremental changes, or changes from one known solution to another. Many researchers have been interested in finding schemes to generate novel solutions, and a genetic approach seems promising because that appears to be the way that novelty has been introduced in nature too.

What then are the basic principles and steps of the genetic algorithm (GA)? There are three aspects to the scheme, mirroring the natural selection process, i.e. (i) survival, (ii) reproduction/crossover and (iii) mutation. In addition, there are analogies between the 'phenotype' and the physical solution on the one hand; as well as the 'genotype' and a low level representation of the solution (generally as binary strings) on the other. The binary string comprises concatenated (i.e. joined up) bit strings, each of which represents a particular state of a variable. So if a variable has 8 discrete states that need to be checked, each state can be represented by a binary number that ranges from 000 to 111; only 3 bits are required to represent the 8 states of that variable. If there are 3 such variables whose values define the solution, each of the variables with only 8 discrete values, then an entire solution can be represented by a ($3 \times 3 =$) 9 bit binary string comprising the three separate bit strings for each variable. The trial solution that is represented by such a concatenated string can be tested with respect to an objective—e.g. minimizing the self-weight of a load carrying structure defined by the values of the 3 variables. These strings correspond to the genotypes in Darwinian evolution.

We could start off with a given population of strings, say around 50 for example, with the bits of the strings (i.e. the ones and zeros) assigned purely randomly. The solutions that are represented by each string can then be evaluated, after mapping the binary states back to the variable space. The trial solutions correspond to the phenotypes (or species) in Darwinian evolution. We can now rate the 50 solutions with respect to their 'fitness'. If this is a weight minimization problem, small values of the objective function (i.e. weight of a given solution) will be assigned a higher fitness—this can be done in some arbitrary way, but desirable solutions must be assigned with higher numerical values of fitness. Constraints will also need to be imposed—e.g. to ensure that a solution does not violate stress limits. We then have to ensure that the phenotypes 'survive' in proportion to their fitness. This should be done in a probabilistic fashion, for example by using a 'weighted roulette wheel' (Goldberg 1989). Phenotypes with larger fitness will probably be chosen many times, while those with smaller fitness may not be chosen at all. The total population size must remain the same as before—i.e. 50 in our case. This is the GA's version of the 'survival of the fittest'.

The next process is 'reproduction' or 'crossover', this time at the level of the genotype, where parts of strings are combined with those of others. This could cause better candidates to be generated than merely the strings that survived the fitness test. In design, it could cause novel solutions to be generated. The reproduction or crossover process is carried out by choosing partner strings for mating; selecting a point in the string for splitting the strings; and combining the first part of one string with the second part of its 'partner'. The choice of partners for 'mating' and the point of crossover are chosen randomly. It should be noted that only the strings corresponding to solutions which have high fitness in the first generation are allowed to reproduce. Crossover would hence result in a new and hopefully fitter population of strings. Mutation is a process (again effected randomly but with a much lower frequency) where a single bit is changed—i.e. from 1 to 0 or vice versa.

The above process is allowed to continue for many generations. The end point of the process is considered to be when incremental improvement in fitness becomes small. This may be slightly inadequate for a strict definition of searching for an optimum, but would suffice for most practical engineering purposes. GAs combine randomness (in the probabilistic rules) with directedness (through the fitness function). It is this combination that makes the scheme analogous to Darwinian evolution, where random variations in the genotype give rise to creative solutions, while the harsh environment ensures error elimination (see Table 3.1). Two good examples of this method are its use (i) to find a minimum weight truss (Jenkins 1991), which is an example of optimization and (ii) to generate house plans (Rosenman 1997), which is an example of novel design.

It may be argued that Darwinian evolution is 'blind' in that it is not goal directed—as reflected in the title of a book by Richard Dawkins (1986)—whereas the GA has a very definite goal, i.e. the fitness function, to aim at. When considered in this light, evolution and GAs may look conceptually quite different. However, it is argued in some quarters that Darwinian evolution too is characterized by 'convergence' of some sort (Conway Morris 2006), as in GAs. Furthermore, if we treat both processes as ones of competitive adaptation to a harsh environment, through the generation of 'novelty' and the elimination of 'weakness', they can be seen as conceptually similar. They also resonate with Popper's thesis that scientific knowledge advances as its bold conjectures are subjected to critical testing.

3.8 Summary

- Two of Karl Popper's most significant contributions to the philosophy of science were his cyclic problem solving methodology and his focus on falsification.
- Engineering processes can be seen as cyclic in nature, whether in a design project or in the wider growth of engineering knowledge. These processes either envisage failure modes or learn from examples of failure. As engineers we can look to Popper for a philosophical underpinning of such processes and approaches.
- Envisaging failure modes in the *design cycle* leads to better design solutions. In the *world cycle* however, we can improve our calculation procedure models, not by actively seeking real world failures, but by learning from failures when they do occur.
- An examination of historical real world failures can help us to identify which component of the calculation procedure model has been falsified—i.e. whether it is the engineering science core; or the outer shells of idealization, margins of safety, design philosophy, design context or engineering process.

– Two relatively recent problem solving approaches also fall squarely within this cyclic paradigm. One of them, the genetic algorithm (GA) technique for optimization and design, embodies elements of randomness and error elimination in a cyclic manner. The other, reflective practice systems, encourages us to learn from our experience through reflection at progressively deeper levels. Above all, we must see engineering and our involvement in it as a learning experience, leading to continuous improvement.

Acknowledgements Adapted from *The Structural Engineer 85*(2), 32–37, Engineering as cyclic problem solving—some insights from Karl Popper by W. P. S. Dias, 2007, published by the Institution of Structural Engineers, London.

References

C. Alexander, *Notes on the Synthesis of Form* (Harvard University Press, Cambridge, 1964)

N. Anwar, *Structural Design Review: Bowling and Theater Roof Truss—Central Rama III* (ACECOMS, Asian Institute of Technology, Bangkok, 1997)

A.W. Beeby, Partial safety factors for reinforcement. Struct. Eng. **72**(20), 341–343 (1994)

D.I. Blockley, *The Nature of Structural Design and Safety* (Ellis Horwood, Chichester, 1980)

D.I. Blockley, Engineering from reflective practice. Res. Eng. Des. **4**, 13–22 (1992)

D.I. Blockley, J.R. Henderson, Structural failures and the growth of engineering knowledge. Proc. Inst. Civ. Eng., Part 1 **68**, 719–728 (1980)

BS 8110: 1997. *Structural Use of Concrete* (British Standards Institution, London, 1997)

S. Cammelli, Tianjin CTF financial centre: wind, form and structure. Struct. Eng. **96**(9), 14–21 (2018)

P. Checkland, J. Scholes, *Soft Systems Methodology in Action* (Wiley, Chichester, 1990)

S. Conway Morris, Darwin's compass: how evolution discovers the song of creation (The Boyle Lecture 2005). Sci. Christ. Belief **18**(1), 5–22 (2006)

R. Corvi, *An Introduction to the Thought of Karl Popper* (Routledge, London, 1997)

R.D. Coyne, M.A. Rosenman, A.D. Radford, M. Balachandran, J.S. Gero, *Knowledge Based Design Systems* (Addison-Wesley, Reading, 1990)

CP 110: 1972. *The Structural Use of Concrete* (British Standards Institution, London, 1972)

R. Dawkins, *The Blind Watchmaker* (Longman, London, 1986)

N.J. Delatte, *Beyond Failure: Forensic Case Studies for Civil Engineers* (ASCE Press, Reston, 2009)

W.P.S. Dias, Structural failures and design philosophy. Struct. Eng. **72**(2), 25–29 (1994)

W.P.S. Dias, Reflective practice, artificial intelligence and engineering design: common trends and inter-relationships. Artif. Intell. Eng. Des. Anal. Manuf. (AIEDAM) **16**, 261–271 (2002)

W.P.S. Dias, Engineering as cyclic problem solving—some insights from Karl Popper. Struct. Eng. **85**(2), 32–37 (2007)

P. Dias, The disciplines of engineering and history: some common ground. Sci. Eng. Ethics **20**(2), 539–549 (2014)

W.P.S. Dias, D.I. Blockley, Discussion on "Galileo's confirmation of a false hypothesis: a paradigm of logical error in design by Henry Petroski". Civ. Eng. Syst. **11**, 75–77 (1994)

W.P.S. Dias, D.I. Blockley, Reflective practice in engineering design. ICE Proc. Civ. Eng. **108**(4), 160–168 (1995)

W.P.S. Dias, U.A. Padukka, AI techniques for preliminary design decisions on column spacing and sizing. Paper presented at the 8th international conference on the application of artificial intelligence to civil, structural and environmental engineering, Rome, 30 Aug–2 Sep 2005 (2005)

W.P.S. Dias, A.D.C. Jayanandana, M.C.M. Fonseka, A.A.D.A.J. Perera, Distress in prestressed concrete roof girders at cement plant. ASCE J. Perform. Constr. Facil. **8**(1), 6–15 (1994)

P. Dias, R. Dissanayake, R. Chandratilake, Lessons learned from tsunami damage in Sri Lanka. ICE Proc. Civ. Eng. **159**, 74–81 (2006)

P. Feyerabend, *Against Method: Outline of an Anarchistic Theory of Knowledge* (New Left Books, London, 1975)

D.E. Goldberg, *Genetic Algorithms in Search, Optimisation and Machine Learning* (Addison-Wesley, New York, 1989)

J.H. Holland, *Adaptation in Natural and Artificial Systems* (University of Michigan Press, Ann Arbor, 1975)

I. Hybs, J.S. Gero, An evolutionary process model of design. Des. Stud. **13**(3), 273–290 (1992)

W.M. Jenkins, Structural optimization with the genetic algorithm. Struct. Eng. **69**(24), 418–422 (1991)

J. LeMasurier, D.I. Blockley, D. Muir Wood, An observational model for managing risk. ICE Proc. Civ. Eng. **159**(6), 35–40 (2006)

M. Levy, M. Salvadori, *Why Buildings Fall Down* (W.W. Norton & Co., New York, 1992)

P. Lipton, The truth about science (The Medawar Lecture 2004). Philos. Trans. R. Soc. Lond. B **360**, 1259–1269 (2005)

I.A. MacLeod, A strategy for the use of computers in structural engineering. Struct. Eng. **73**(21), 366–370 (1995)

B. Magee, *Popper* (Fontana, London, 1973)

M.A. Notturno, *Science and the Open Society: The Future of Karl Popper's Philosophy* (Central European University Press, Budapest, 2000)

H. Petroski, *To Engineer Is Human: The Role of Failure in Successful Design* (St. Martin's Press, New York, 1969)

H. Petroski, Galileo's confirmation of a false hypothesis: a paradigm of logical error in design. Civ. Eng. Syst. **9**(3), 251–263 (1992)

H. Petroski, *Design Paradigms: Case Histories of Error and Judgment in Engineering* (Cambridge University Press, Cambridge, 1994)

N.F. Pidgeon, D.I. Blockley, B.A. Turner, Design practice and snow loading—lessons from a roof collapse. Struct. Eng. **64A**(3), 67–71 (1986)

K.R. Popper, *The Poverty of Historicism*, 2nd edn. (Routledge and Kegan Paul, London, 1960)

K.R. Popper, *The Logic of Scientific Discovery*, 2nd edn. (Hutchison, London, 1968)

K.R. Popper, *Objective Knowledge: An Evolutionary Approach* (Oxford University Press, Oxford, 1972)

K.R. Popper, in *Realism and the Aim of Science: Postscript to the Logic of Scientific Discovery*, vol. 1, ed. by W.W. Bartley III (Hutchison, London, 1983)

K.R. Popper, *Conjectures and Refutations: The Growth of Scientific Knowledge*, 5th edn. (Routledge, London, 1989)

K.R. Popper, *All Life is Problem Solving* (Routledge, London, 1999)

M.A. Rosenman, An exploration into evolutionary models for non-routine design. Artif. Intell. Eng. **11**, 287–293 (1997)

D.A. Schon, *The Reflective Practitioner: How Professionals Think in Action* (Temple Smith, London, 1983)

P.M. Senge, *The Fifth Discipline: The Art and Practice of the Learning Organization* (Century Business, New York, 1992)

P.G. Sibly, A.C. Walker, Structural accidents and their causes. Proc., Inst. Civ. Eng., Part 1 **62**, 191–208 (1977)

Y. Umeda, H. Takeda, H. Yoshikawa, T. Tomiyama, Function, behaviour and structure, in *Applications of Artificial Intelligence in Engineering V, Vol. 1—Design*, ed. by J.S. Gero (Computational Mechanics Publications, Southampton, 1990), pp. 177–193

W.G. Vincenti, *What Engineers Know and How They Know It: Analytical Studies from Aeronautical History* (Johns Hopkins, Baltimore, 1990)
S.B. Willoughby, The Ridgeway footbridge. Struct. Eng. **74**(5), 79–83 (1996)

Chapter 4
Will Any Old Model Do?

4.1 Paradigms and Revolutions

Thomas Kuhn's (1922–96) philosophy of science arose from the historical investigations he conducted regarding the progress of science. Those investigations led him to propose the idea that science progressed through significant *revolutions* that occurred from time to time—an example he uses frequently is the Copernican revolution that placed the sun, and not the earth, at the centre of the solar system. In between such unsettling periods of change, *normal science* was conducted within what he called a 'ruling paradigm'. He also described the way that *paradigm shifts* occurred during revolutions.

One of the main questions raised by Kuhn's philosophy is the extent to which the practice and progress of science is influenced by psychological, sociological, cultural and historical forces. In other words, he asked whether it was culture or nature that determined the shape of science (Weinberg 1998). He also suggested that science did not progress *towards* the 'truth' through its revolutions, but only *away from* its primitive beginnings. After Kuhn, others have moved to rather extreme positions, which hold that scientific knowledge is only a 'social construct', i.e. merely a product of sociological influences and not a true description of nature (Barnes and Bloor 1982). We shall examine how relevant these ideas are to engineering (see also Dias 2008).

4.2 Normal Science Within the Ruling Paradigm

The main sense in which Kuhn used the term 'paradigm' was to denote the shared commitments of a scientific community. He later preferred the term 'disciplinary matrix' to describe this. A disciplinary matrix would certainly include a common education; a body of literature that was commonly read; and a shared acceptance of methods and equipment that are deemed to be appropriate for tackling problems,

P. Dias, *Philosophy for Engineering*, SpringerBriefs in Applied Sciences and Technology, https://doi.org/10.1007/978-981-15-1271-1_4

which in turn should only be 'admissible' ones. It could also involve metaphysical factors, and other guiding principles such as 'predictions should be accurate', 'quantitative is better than qualitative' and 'the principle of parsimony'.

Kuhn went on to describe the way that scientists were educated, and suggested that there was close correspondence between such education and the practice of normal science within the ruling paradigm. The student was trained essentially on text books, which contained bodies of *established* knowledge. Scientists were not exposed to cutting edge knowledge (in journals for example) until very late in their training. In addition, science text books contained little if any descriptions of the historical development of their subject matter. Paradigms or frameworks that operated previously were not considered important; only the prevailing one was (Kuhn 1970). There was very little scope for teaching students to discriminate between different points of view, because all scientific text books in a given field had the same point of view—i.e. the one based on the ruling paradigm (Kuhn 1977). The solving of (generally numerical) problems after studying topics in a text book was also typical of such education. This was an exercise in training the budding scientist to apply established knowledge to new problem areas (Kuhn 1977).

The objective was to form the students into a very definite mould—i.e. the ruling paradigm. Such education has been described by Kuhn (1970, p. 166) as "a narrow and rigid education, probably more so than any other except perhaps in orthodox theology". This led to a mode of thinking that would be called *convergent. Divergent* thinking could be important for architects, inventors, artists and philosophers, since it promoted creativity. Kuhn argued however, that scientists had to be convergent thinkers, since most of them had to work within a given paradigm, during long periods of normal science (Kuhn 1977).

A scientific education also promoted a spirit of dogmatism, which was required for the practice of normal science. Words such as 'faith', 'trust' and 'taken for granted' figure in Kuhn's writings (Hoyningen-Huene 1993); such words are more usually associated with religious faith. However, Kuhn argued that such faith and dogmatism created the background within which error or anomaly could 'stick out' (Kuhn 1977) and lead to a change in the paradigm. The very rigidity of the tradition ensured the shattering of that selfsame tradition. He endorsed Bacon's maxim that "truth emerges more readily from error than from confusion" (Kuhn 1970), in preferring a single paradigm to govern normal science, rather than to have several simultaneous alternatives, as preferred by other philosophers of science (Feyerabend 1981; Lakatos 1981).

4.3 Scientific Revolutions and Progress

4.3.1 The Nature of Revolutions

Kuhn (1970) used the metaphor of a political revolution to describe a scientific revolution that occurred from time to time. Just as a political revolution arose out of growing *discontent*, a scientific one did so out of growing *anomaly*. A growing number of anomalies was considered to constitute a *crisis*, just as in a political revolution. Kuhn used the word 'anomaly' in a fairly specific way to denote unexpected discovery and 'crisis' to suggest mismatches between theory and observation. The mere existence of an anomaly (however significant) did not lead to a paradigm shift. In order for that to happen, there had to be a competing paradigm that would explain the anomaly, as well as everything else the previous paradigm explained. If such an alternative was not available, the existing paradigm would be modified, even by some ad hoc measures.

During a period of crisis, the anomalies became the focus of research within the scientific community. Many versions or interpretations of the paradigm were proposed, together with a willingness to 'try anything' and work outside the existing paradigm. There were debates over fundamentals that had always been part of the paradigm and taken for granted during the period of normal science. There was even a turning to metaphysics and philosophy. In short, a period of crisis was one of great insecurity. Such a crisis was described by Kuhn as the signal to move from a period of normal science to a scientific revolution. Just as in a political revolution, a new order was brought into existence, with considerable disjunction from the previous one, although adherents to the old order continued for a while. History was also re-written in some cases, from the viewpoint of the new order (Kuhn 1970).

4.3.2 A Different World

Kuhn said that "though the world does not change with a change of paradigm, the scientist afterwards works in a different world" (Kuhn 1970, p. 121). There were many ways of understanding this. First, scientists acquired different definitions of words that relate to the world, and pursued different problems too; this could require new equipment and methodology as well. For example, a number of new planets and comets were discovered after the Copernican Revolution, because scientists were looking for them; a number of new types of rays were also discovered after Roentgen's discovery of X-rays (Kuhn 1977).

Next, the world could also be perceived differently. To Galileo, a stone at the end of a chain was a pendulum (which has the same period whatever the amplitude), whereas Aristotle had seen it as a 'constrained fall'—i.e. the fall of the stone to the ground was constrained by the chain. The measurements considered important by Galileo were very different to those made (if any) by Aristotle. Galileo's perception was very 'fruitful', leading as it did to other problems being solved by using the same principle of potential energy conversion to kinetic energy—e.g. flow through an orifice under a falling head of water.

Finally, because data was always 'seen' through the eyes of theory, a perceptual change could actually cause a change in data. Consider Kuhn's account of the change in chemical composition of compounds after Dalton's atomic theory that atoms combined in simple whole-number ratios (Kuhn 1970, pp. 134–5):

> When Dalton first searched the chemical literature for data to support his physical theory, he found some records of reactions that fitted, but he can scarcely have avoided finding others that did not. Proust's own measurements on the two oxides of copper yielded, for example, an oxygen weight-ratio of 1.47:1 rather than the 2:1 demanded by the atomic theory; and Proust is just the man who might have been expected to achieve the Daltonian ratio. He was, that is, a fine experimentalist, and his view of the relation between mixtures and compounds was very close to Dalton's. But it is hard to make nature fit a paradigm. That is why the puzzles of normal science are so challenging and also why measurements undertaken without a paradigm so seldom lead to any conclusions at all. Chemists could not, therefore, simply accept Dalton's theory on the evidence, for much of that was still negative. Instead, even after accepting the theory, they had still to *beat nature into line*, a process which, in the event, took almost another generation. When it was done, even the percentage composition of well-known compounds was different. The *data themselves had changed*. (Italics by present author)

It must be emphasized that Kuhn distinguished between interpretation and perception, saying that it was *perception* that changed during a revolution, and not merely interpretation. This has been likened to a 'Gestalt-shift' that enables Fig. 4.1 to be viewed as either a duck or a rabbit. Kuhn however said that perceptual changes were permanent, and compared them to a person wearing inverting spectacles—once the change was made, it was difficult to see the world in the previous way (Kuhn 1970).

The above idea of the world is what has led to Kuhn's ideas being used by some—mostly social scientists rather than natural scientists—to suggest that science is purely a social construct, and not grounded in any reality or truth of nature (Barnes 1982). Hoyningen-Huene's (1993) interpretation of Kuhn's philosophy is however that the 'world-in-itself' always shows 'resistance' to socially constructed paradigms. It does so by raising anomalies that make paradigm changes necessary—and in many cases such anomalies occur simultaneously among different groups of scientists; these

Fig. 4.1 "Duck-Rabbit" image that produces "Gestalt-switching" (from Dias 2008)

anomalies have historically been explained by science, together with the previously explainable phenomena, by a better paradigm. In other words, although a paradigm cannot fully encapsulate nature—because much of it would still be unknown—it is certainly governed by it.

4.3.3 Progress Through Revolutions

We have described the progress *of* a revolution. Kuhn also made some comments as to what progress in science was made *through* a revolution, and generated much controversy in the process. According to Kuhn, science did not progress *towards* 'truth', but rather *away from* its primitive beginnings. He drew parallels between this idea and that of Darwinian evolution, which was based on survival, and not goal directed (Kuhn 1970). This is diametrically opposed to Popper's idea of truth as a goal that we kept aiming for (Blockley 1980). Most scientists too would reject the idea that they were not getting closer to a correspondence with nature through their theories (Weinberg 1998).

Kuhn's thoughts regarding progress were probably based on his ideas that scientific revolutions were essentially changes in a scientific community's perceptions of the world; also that such perceptions were socially conditioned to a considerable extent. Furthermore, he gave a number of examples where elements of a previously overthrown paradigm were 'resurrected' after a subsequent revolution (see Table 4.1). It must be noted of course that the 'resurrected' forms differed from their previously 'buried' ones, not only in form, but also in sophistication and accuracy. Kuhn (1977) also acknowledged that scientific revolutions always resulted in greater quantitative accuracy, and that our ability to predict phenomena had therefore increased steadily with time.

Returning to Table 4.1, we see that Aristotelian science was governed by the idea that objects and materials had certain 'essences' which determined their behavior, including motion—so a stone fell to the earth because that was its nature or 'essence'. This notion was considered too shallow during the scientific revolution, especially to explain motion, and even before Newton, it had been replaced by the 'corpuscular' theory of matter—i.e. that the properties of matter could be explained by the

Table 4.1 Examples of some reversals in paradigm shifts (from Dias 2008)

Subject area	Original idea	Changed idea	Reversed idea
Nature of matter	'Essential' properties (Aristotelian)	Corpuscles (pre-Newtonian)	'Innate' properties (Newtonian)
Nature of space	'Place' has potency (Aristotelian)	'Space' is referential (Newtonian)	'Space-time' is curved (Einsteinian)
Nature of universe	Heliocentric universe (Aristarchus)	Geocentric universe (Aristotle)	Heliocentric solar system (Copernicus)

interaction of particles—so the reason we could smell substances was because they emanated particles that could reach our noses. When Newton had to explain gravity however, he had to revert back to the idea of 'innate' properties of matter (since 'particles' such as celestial bodies could not interact at great distances), an idea very close to Aristotle's 'essences'. This idea of 'innate' properties was also very useful in the subsequent development of chemistry and electricity (Kuhn 1970).

Another example was Aristotle's notion of 'place', which was supposed to exert an influence on objects—so another reason a stone fell towards the earth was because of the earth's place at the centre of the universe. This was replaced by Newton's idea of space as a 'frame of reference' that had no influence on objects. However, Einstein's space-time continuum is both influenced by the mass of celestial bodies and also influences the path of light rays. Hence Aristotle's notion of space could be considered closer to Einstein's in some respects than Newton's was (Kuhn 1985). Probably the clearest example of paradigm reversal is that Aristarchus, a pre-Socratic philosopher, proposed a heliocentric universe in the 3rd century BC (Kuhn 1985). This theory of course 'lost out' to Aristotle's geocentric universe, and it was only in the 16th century AD that science returned to heliocentricism, albeit of the solar system and not the universe.

4.4 Revolutions in Structural Design

4.4.1 Some Historical Revolutions

The relevance to engineering of Kuhn's ideas about paradigms and revolutions in science has been cogently presented by William Addis (1990) in his book *Structural Engineering: The Nature of Theory and Design*. Addis does this by using the history of structural design. The design of structures is carried out today largely using the theories of structural mechanics; those theories however are of rather recent origin. Structures have been designed and constructed long before the advent of mechanics. Therefore, they would have had a different basis for their design. A close look at the history of structural design will indicate that there were many different bases for design, and that these changed over the years. In some cases, these different bases co-existed, and still do. This points therefore to the existence of design paradigms. It also raises the question as to how accurate or realistic the various bases for design have been and are for that matter, thus suggesting the relativity of knowledge. The term 'revolution' as used by Kuhn conveys the idea that scientific knowledge is not purely cumulative. Paradigm shifts constitute changes in viewing or perceiving natural and physical phenomena. Similarly, it can be argued that changes in the bases of design constitute changes in the way structures and their behavior have been viewed. We shall now consider some shifts in the ruling paradigm for structural design that have taken place over the centuries, drawing heavily on Addis (1990).

In ancient Greece, the 7th century BC saw a quick transition from small-scale construction to monumental architecture, primarily in the form of temples. The layouts of these temples indicate clearly that some principles of proportion were used in their design. For example, Addis (1990) gives flowcharts for design according to the Doric and Ionic orders, as compiled by Vitruvius in the 1st century BC, where sizing is based largely on proportions. These may have been influenced by harmonic ratios, the discovery of which is attributed to Pythagoras in the 6th century BC. He is credited with finding that small number ratios (e.g. 2:1, 3:2, 4:3 etc.) correspond to the most consonant musical intervals (e.g. octave, fifth, fourth etc.). Anaximander thought that such ratios characterized the structure of the universe, which he considered as analogous to Greek temple architecture (Hahn 2001).

Geometry also played an important part in Greek thought. Its application to structural design was not mathematical however, as in Bow's graphical statics or Heyman's (1982) geometrical factor of safety. Rather, geometry was used to encapsulate harmonic ratios. Furthermore, in the Greek worldview, which was *rational* (i.e. based on reason) but largely not *empirical* (i.e. not based on observation), symmetrical or 'perfect' geometric shapes such as the circle were considered to reflect the nature of the universe. It is therefore not surprising that Roman arches, which followed the Greek period, were all circular ones.

We shall next consider the tremendous developments, primarily in Cathedral architecture, that took place around the 12th century in Europe, characterized by the advent of the Gothic form. In many ways, the bases for these developments were extensions of those used by the ancient Greeks. Geometry also continued to play an important part in the Gothic revolution. It is interesting to note that Adelard of Bath, who was a student of the master builder Thierry at Chartres, translated a copy of Euclid's *Elements of Geometry* from Arabic to Latin just a few years prior to the construction of Chartres Cathedral. However, the use of geometry was now more sophisticated. There was a distinction made between theoretical and practical geometry. Theoretical geometry was concerned with ideal shapes and how structural form could be justified on the basis of those shapes. Practical geometry on the other hand drew on theoretical shapes, but also on experience and thought experiments regarding structural behaviour, in order to derive sometimes complex rules for arriving at appropriate shapes for arches and thicknesses for voussoirs and abutments. Hence, it was possible to move away from a purely circular arch form to pointed and flat arches.

One new basis for design that emerged during this Gothic Revolution was the use of physical scale models. Although their use in most cases may have been architectural as opposed to structural, the 1/8th full size model of the San Petronio Cathedral in Bologna would have constituted a reasonable test of the stability of the proposed cathedral, since the model was made of brick and plaster, materials similar to those to be used in the cathedral itself. Another basis for design that emerged in the Gothic period, simple though it may seem, was the use of precedent in design. In many cases master builders visited other cathedrals prior to commencing their own work. In other cases, the 'client' city consulted master builders who had experience in constructing cathedrals.

The next revolution was the shift from geometry to statics, seen once again in the design of arches. One of the greatest problems with arch design is to carry the horizontal thrust via abutments. In the early 17th century, Blondel proposed a purely geometric approach, which was publicized in a textbook by Belidor in 1729. However, in the middle of the 17th century, Blondel's rule was criticized by Christopher Wren (the designer of St. Paul's Cathedral in London), who argued that principles of statics relating to stability had to be employed in the design of abutments. There were others too who found a lack of rationality in purely geometric approaches. Hence, the geometric design procedure could be seen as having undergone a 'Kuhnian' crisis around the latter half of the 17th century.

Both the geometric and statical approaches were subsumed in the 19th century by the 'elastic design revolution', which was largely concerned with determining the internal stresses and deformations of structural elements. The term 'elastic' conveys the idea that the above stresses and strains are reversible during loading and unloading, with no permanent deformations experienced—so structures are designed to remain within the 'elastic region' during their use. The need to determine internal stresses and deformations arose due to the use of slender elements made of steel and timber that experienced bending and tensile forces in structures such as beams and trusses. Such structures were far more likely to have localized material failures in elements, as opposed to stability failures of the entire structure as in masonry construction. The results of the above stress analyses could also be compared with the increasing body of data that had been acquired in laboratory testing on the properties of materials.

It should be noted that both the geometric and statical approaches were preserved, albeit in transformed fashion, by the elastic design revolution. Statics had to apply to the structure as a whole, in order to ensure equilibrium. It was also applied to parts of the structure as well (e.g. the force equilibrium at a truss joint). The actual application of statical principles to structures was carried out largely by using the graphical (or 'geometric') procedures of the triangle of forces and Bow's notation. The interesting thing about the elastic design revolution is that although it arose in the context of beams and trusses (made of steel and timber), its methods were subsequently applied to all structural forms, including masonry arches. This illustrates Kuhn's proposition that the world (in this case a masonry structure) is viewed in a new light (in this case internal stresses) after a scientific revolution (in this case the elastic design revolution).

4.4.2 The Plastic Design Revolution

The 'plastic design revolution', as masterfully narrated by Addis (1990), contains all the elements of a Kuhnian revolution. Prior to the plastic design revolution, elastic design was the ruling paradigm. Under this earlier paradigm, there was much growth of knowledge in cumulative fashion (as in a Kuhnian period of *normal* science), with the paradigm being applied to an increasing number of situations. The paradigm was supported by advances in structural mechanics and data from laboratory testing. However, the application of elastic theories to indeterminate structures proved to be very tedious and computation-intensive; they could not readily be used for everyday design. As such, many simplifications had to be made.

For example, a steel frame building was seen as a grid of vertical columns, connected by horizontal beams. When designing for vertical loads, the beams were considered to be simply supported at connections to columns; in assessing the loading on columns, some eccentricity in the beam-column connection was allowed for, based on the designer's judgment. On the other hand, when designing for lateral loads, the connections between beams and columns were considered to be fully rigid. Despite the above simplifications and contradictions, steel frame buildings were constructed after being elastically designed for over half a century (in Britain at least), without any recorded instance of collapse. Furthermore, elastic design principles were used and resulted in safe structures of many other forms (e.g. trusses, arches) and materials (e.g. timber, masonry).

An *anomaly* in an existing design paradigm can manifest itself in many ways. One of the most spectacular is through structural failure. Even without failure however, when a paradigm is seen to lack justifying power, an anomaly can be said to have arisen. This is precisely what happened in Britain with regard to the elastic design paradigm. The anomaly arose in, or was focused on the design of steel structures, particularly framed buildings but also bridges. There were many contradictions and deficiencies in the elastic design paradigm. As described above, the beam-column connections were viewed in two different ways when designing for vertical and lateral loads. Furthermore, the eccentricities allowed for were found to be inaccurate when compared with laboratory tests. In addition, the localized elastic stresses due to manufacturing and the internal stresses due to 'lack of fit' during construction could not easily be dealt with using an elastic analysis. In 1929, The British Steelwork Association set up the Steel Structures Research Committee (SSRC) to review the elastic design procedures and to propose design procedures that were more efficient (with respect to design office practice) and economical (with respect to use of construction materials).

The initial work of the SSRC was directed at trying to remedy the above anomalies within the paradigm of elastic design itself. This is typical of a first response to anomaly, according to Kuhn. It resulted in design procedures that were considerably more difficult to follow than the previous ones, with no gain in economy either. This situation, where the remedy was worse than the illness, constituted a *crisis*. There was also the increasing awareness that the use of an 'elastic working stress' (i.e. the maximum possible elastic stress divided by a safety factor) as a design parameter gave wide discrepancies between theoretical and measured values. Furthermore, there was the parallel awareness that elastic stresses gave no information regarding the collapse behaviour of structures. It was also felt that the plasticity of steel (i.e. ability to deform safely even beyond the maximum elastic stress or strain) would constitute reserves of strength that could result in more economical structures.

The crux of the plastic design *revolution* was the shift from calculating elastic stresses in structural elements while in service, to determining loads that structures could carry at the point of collapse. There were many new concepts that had to be introduced, such as 'perfect plasticity', 'plastic hinges', 'load factors' and 'collapse loads'. The focus of attention too changed from the service state of a structure to its ultimate or collapse state. The technique of 'superposition', used for elastic design, was replaced with that of 'proportional loading'. New theorems such as the safe and unsafe theorems of plastic theory were developed. There were many advantages that resulted from the revolution. It was found that predicting the actual (i.e. measured) collapse loads by plastic theory was far more accurate than predicting measured working stresses by elastic theory. The plastic theory solutions were also not as sensitive to variations in dimensional and connection details as were the elastic ones. Hence, the new paradigm had greater justifying ability. Furthermore, as a result of utilizing the plasticity of the material and the redundancy of the structure, greater reserves of structural strength were available for the designer, thus leading to economical design. The plastic design methods were not tedious either, and could deal with redundancy very easily via the concept of plastic hinges. All of the above contributed towards the gradual acceptance of the new plastic design paradigm over that of the elastic design one.

Articulation of the new paradigm was carried out in many ways. For one thing, the paradigm was applied to many other design procedures as well. Today, structures constructed out of virtually all construction materials are designed according to plastic (or 'ultimate limit state') design principles. For another, there was a 're-writing of history' as observed by Kuhn, when work done by Coulomb and others in the 18th century that bore some similarity to the plastic design paradigm was recast by Heyman in the categories of the new paradigm. There was also an interesting 'return to history' in the case of the plastic design procedures for masonry arches, since the importance of geometry was re-emphasized by the new paradigm over that of stress analysis (Heyman 1982). Finally, the new paradigm found articulation in a new design community, trained largely in some universities that had their academic staff members in the SSRC. It should be noted that the elastic design community too continued, with periodic academic debate regarding the suitability of the rival paradigm. This is reminiscent of the 'rearguard' action, described by Kuhn, by adherents to the overthrown paradigm. It must be said however that the plastic design paradigm has rapidly gained acceptance because of its greater justifying ability, its simplicity and robustness, and its capacity to arrive at more economical structures.

4.4.3 Relativity and Progress of Knowledge

The design revolutions described above all indicate that knowledge is relative, in the sense that we begin to 'see' structures and their behaviour in different ways after such revolutions. This however is not to say that each way of seeing things was or is as good as the other. To take such a position of randomness, such as advocated by Feyerabend (1981), would be anarchic. Where structural design is concerned, the design revolutions over the years have certainly contributed to progress in knowledge, particularly with regard to greater justifying ability. We could also say that the models are able to pass an increasing number of 'tests'—i.e. the truth contents of our models have increased. They have also given us greater confidence and in some cases improved quantitative accuracy, thus enabling us to construct slender and more economical structures.

It may be that sometimes a ruling paradigm returns to a previous one; as in the case of masonry arch design, when the plastic design revolution rekindled a geometric approach to design, after it had previously been ousted by the stress analysis approach of the elastic design paradigm. Nevertheless, such a return always takes place with increased understanding and fresh perspectives. Hardly ever have there been simple and complete returns to previous approaches. The question could then be asked as to how close our models can get to the 'real world'. As we shall see in the next section, such perfect correspondence, apart from being practically impossible and philosophically futile according to Kuhn, is not the primary concern of engineering models either.

4.5 Engineering Models

The models used by engineers could be considered as part of their paradigm. We shall compare engineering *models* with scientific *theories*, following Blockley (1992). Although the idea of a model is used in science too, the term there refers to some sort of analogy—for example a fluid flow model for an electrical circuit. Engineering models are primarily calculation procedure models (CPMs) (Sects. 3.4 and 3.5)—i.e. tools for making changes in the world, incorporating both scientific and heuristic aspects. Now both an engineering model and a scientific theory are *representations* of the world. They are both used to make *predictions* about the world. How then do they differ?

The discussion below is summarized in Table 4.2. The goal of science is *understanding*, while that of engineering is *transformation* or useful change (see Fig. 2.1).

Table 4.2 Scientific theories compared with engineering models (after Dias 2008)

Feature	Scientific theory	Engineering model
Goal	Understanding	Transformation
Grounding	Truth	Safety
Basis	Necessity	Contingency
Form	Simplicity	Completeness
Applicability	Comprehensiveness	Practicality
Specification	Precision	Appropriateness
Improved by	Calibration	Comparison
Characteristic	Accuracy	Dependability

So, while both scientific theories and engineering models represent and make predictions about the world, their purposes are very different, as unfolded in the rest of the table. For example, we can say that a scientific theory is grounded in *truth*, in the sense that a good theory has a good correspondence with the world (or natural phenomena). Hence, the basis of a scientific theory is *necessity*, in that the theory cannot do anything other than to reflect the laws of nature (Goldman 2004).

An engineering model is however grounded more in *safety*, in that a good model will help us to make safe artefacts or objects. This is why Kuhn's view of science is probably more relevant for engineers than for scientists. In the various historical stages of structural design, structures were viewed through the 'eyes' of proportions, geometry, statics, elasticity and plasticity. While there has undoubtedly been an increase in the truth content of engineering models with time, what is perhaps more important is that each historical stage was served by its own model or paradigm for the purpose of constructing safe structures. Furthermore, the basis of an engineering model used for designing useful things is *contingency*—while the design has to respect the laws of nature, it is influenced by a variety of context dependent factors that have to be recognized and accounted for (Goldman 2004).

This pragmatic engineering approach is reflected in the aphorism that "it's true because it works"; whereas scientists are probably more interested in claiming for their theories that "it works because it's true". It also explains why engineers are reluctant to make changes in a CPM once it has been shown to result in safe artefacts—"if it ain't broke, don't fix it"—however 'unjustifiable' parts of that CPM may be. This pragmatism is also reflected in the form, applicability and specification of engineering models.

Scientific theories aim at *simplicity*. One of the best expositions of this quality has been made by Einstein (1934) himself, in an address delivered in 1918 at a celebration of Max Planck's sixtieth birthday:

> What place does the theoretical physicist's picture of the world occupy among all these possible pictures? It demands the highest possible standard of *rigorous precision* in the description of relations, such as only the use of mathematical language can give. In regard to his subject matter, on the other hand, the physicist has to limit himself very severely: he must content himself with describing the *most simple* events which can be brought within the domain of our experience; all events of a more complex order are beyond the power of the human intellect to reconstruct with the subtle accuracy and logical perfection which the theoretical physicist demands. Supreme purity, clarity and certainty at the cost of *completeness*. But what can be the attraction of getting to know such a tiny section of nature thoroughly, while one leaves everything subtler and complex shyly and timidly alone? Does the product of such a modest effort deserve to be called by the proud name of a theory of the universe? In my belief the name is justified: for the general laws on which the structure of theoretical physics is based claim to be valid for any natural phenomenon whatsoever. (Italics by present author)

Compare this with the importance given to handling incompleteness in engineering. While simplicity may be important in scientific theories for clarity of explanation, engineering models by contrast focus on *completeness*. Consider for example the part plan view in Fig. 4.2 of a series of combined footings on weak soil connecting two lines of peripheral columns. The foundation design for the typical 'combined

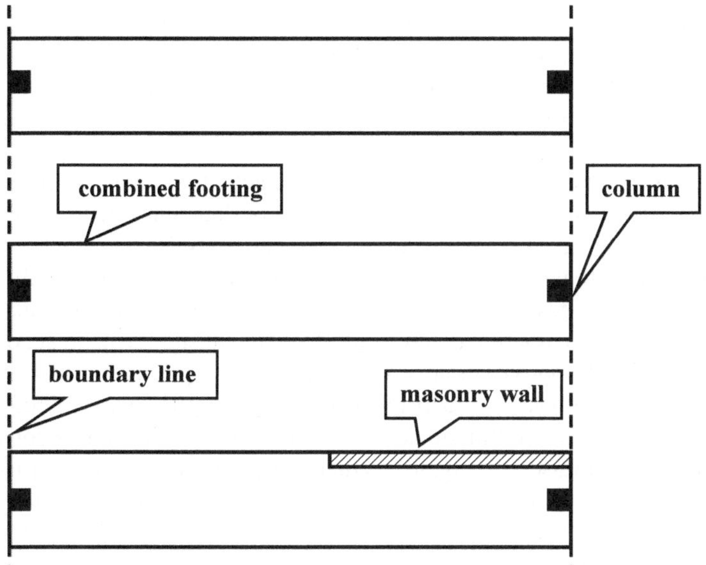

Fig. 4.2 Part plan view of combined footings supporting two lines of boundary columns (from Dias 2008)

footings' are simple, and standard engineering science solutions will suffice. One footing however, has an 'inconvenient' wall segment founded on it too. Such situations abound in the practice of engineering, because it is contingent on context. The artefacts to be designed cannot be fitted neatly into standard solutions. There are peculiarities that have to be dealt with—they cannot be ignored. This is one aspect of completeness. The notion of *uncertainty*, which is part of everyday life, leads us to another aspect of completeness that needs to be tackled by engineering models. There are three aspects of uncertainty, namely randomness, fuzziness and incompleteness (Blockley 2013); they have been addressed in Sect. 2.3

We come next to applicability. For scientific theories, *comprehensiveness* is the important aspect of applicability. A scientist (perhaps an engineering scientist) looking at Fig. 4.2 would have his eye attracted to the typical footings. He may then think of writing a spreadsheet program that can solve such foundation problems for any combination of column load, footing span and soil bearing pressure. This would of course be of considerable value. But it does not solve the practical engineer's problem that one of the footings has an eccentric wall load along part of its length. For engineering models therefore, *practicality* is the relevant aspect of applicability.

How about specification? As stated by Einstein above, scientists would aim for *precision* in their predictions. On the other hand, the important thing for engineering models is *appropriateness*. Consider again the atypical footing in Fig. 4.2. Our (engineering) scientist would want to analyze this precisely (e.g. locating the wall in its exact location), perhaps by using some computerized analysis program. There are a number of questions that arise in this regard. First, is it appropriate to spend the time and effort required for very precise modelling to solve a relatively inconsequential problem such as this (given that the masonry wall loads would be much less than those from the columns)? Second, is it of any value to use such precise analysis when the properties of the underlying soil could be quite variable and known only approximately? Third, if we aim at such precision, will we be in danger of getting too close to the point of failure, especially if the actual parameters differ from the assumed ones (due either to variability in the soil or deficiency in construction)?

A practising engineer may prefer to use a cruder analysis, assuming for example that the wall runs along the entire length of the footing; or that such wall segments are present on both edges of the footing (so that some symmetry is introduced). More than one simple approximation could be used and the design made to accommodate the worst cases. Such approaches are probably more appropriate, and characteristic of engineering. For this however, practitioners must have a qualitative feel for structural behaviour, so that the approximations used are more as opposed to less conservative.

The other insight raised by Fig. 4.2 is that its peculiarities are part of its *context*, something that is crucial for the practice of engineering. Context generates *constraints*, or limits what we can do. While they appear to be curtailing our freedoms, they in fact give us challenges to face and also provide a 'solution space' within which to work—and hence unleash creativity. So, although the location of the masonry wall in Fig. 4.2 complicates our calculations a little, the final solution is elegant and efficient, in that the wall is founded on the combined footing itself, dispensing with any need for a separate foundation of its own.

The penultimate row in Table 4.2 indicates that scientific theories are improved by *calibration* with the world, whereas engineering models are improved by *comparison*. Calibration is relatively easy when theories are simple and pay 'selective inattention' to factors that cannot be easily quantified (Schon 1983); provided that the experiment or observation that is used for calibration is 'freed' or 'decontextualized' from such factors too—see the Einstein quotation above. Engineering however is carried out within a complex social fabric, and such decontextualization is neither possible nor even desirable. Creative engineers have to pit their wits against the cunning adversary of nature and the indeterminism of human action. However, this means that their models cannot be calibrated as such against the world. Comparison is a better word to use for the feedback that engineering models receive from completed artefacts. If there are problems or failures that are revealed from such comparisons, engineers seek to remedy the appropriate component of their CPM (Dias 1994)—see also Sect. 3.6.

Most of the differences above stem from those expressed in the first two rows of Table 4.2, and this results in the distinctive characteristics of scientific theories and engineering models (last row of table). Because scientists seek to understand the world, their goal is truth. Engineers on the other hand are interested in changing the world, and want to do it safely. The concept of safety involves an asymmetry between predictions that are to the two different sides of a 'target' that the concept of truth does not—see Fig. 4.3. So, a scientist predicting a lunar eclipse will want to get the time as accurately as possible. There is no practical difference if her prediction was slightly earlier or later than the actual time of the eclipse; she does however want to be

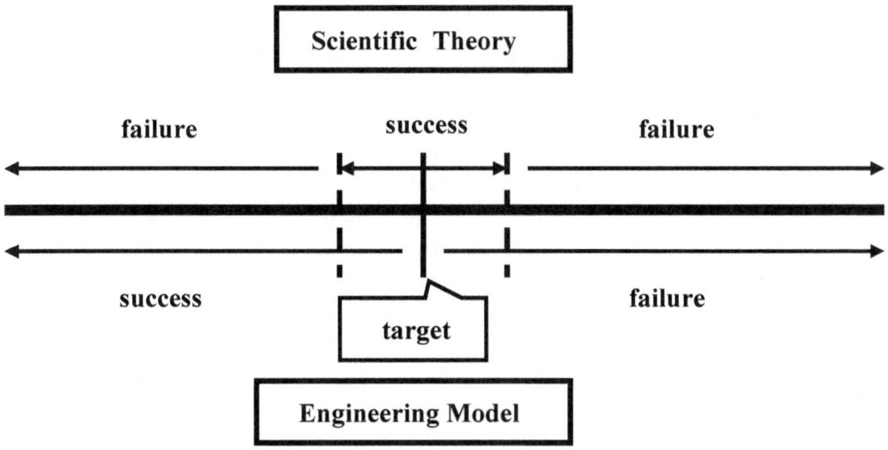

Fig. 4.3 Success and failure for predictions in science and engineering (from Dias 2008)

as close as possible to the actual time and will therefore aim for the highest *accuracy*. For an engineer on the other hand, there is a large difference between underestimating the stability of a structure or reliability of a system and overestimating it. The latter would cause failure and violate his pursuit of safety. This is why engineering models tend always to be underestimates of safety; an engineer requires *dependability*.

This also explains why very different theories, ranging from proportions to plastic analysis, were satisfactory for constructing structures. It demonstrates too that engineers, like social scientists, can more easily accommodate the relativity of knowledge suggested by Kuhn than can natural scientists. Note also that codes of practice, which formalize design procedures, despite varying from country to country sometimes in very fundamental ways, still manage to produce safe structures and artefacts in those different countries. This is another example of how engineering CPMs, of which codes of practice are an important part, reflect the above relativity of knowledge. It must be noted however, that concerns regarding economy will tend to push engineers closer to the 'target' in Fig. 4.3—we want safe structures, but not at unreasonable cost.

Engineers are drilled in engineering science during their formal education, because engineering is founded on science. In such courses they become used to getting a single correct answer or a *unique* solution. They are however given a good dose of design too in an engineering first degree. Such courses in design emphasize that engineering models should be characterized by *dependability* rather than accuracy. In other words, solutions must have generous *margins of safety*, not only in the design assumptions and calculation procedures, but also in the selection of structural forms that can, through their robustness, accommodate unforeseen actions.

Design is also founded on creativity and an artistic outlook. These require divergent thinking. Engineers cannot be divergent thinkers when it comes to engineering science—all our structural solutions are constrained by Newton's laws and energy theorems. However their creativity can be unleashed in the selection of structural form for both robustness and aesthetics. This could involve a return to previous design paradigms. For example Bobrowski (1982) was drawn to the stability of a circle (*geometry*) in developing the plan form for his Calgary Saddledome. Some work has also been done on the importance of *proportions* for creating aesthetic forms (Kulasuriya et al. 2002). So, although history is not important for scientists, as argued by Kuhn, structural engineers can and should draw on historical paradigms when appropriate. It is this recourse to history and to nature as well, that will stimulate creativity and generate a pool of *alternatives*, as opposed to a unique solution.

4.6 Summary

- Thomas Kuhn's investigations on the history of science led him to conclude that scientists worked within a 'ruling paradigm' during relatively long periods of 'normal science'. From time to time however this paradigm could be replaced by another through the process of a 'scientific revolution'.
- The history of structural design has been presented by William Addis as one of changing paradigms, with the bases of design moving from proportions though geometry, statics and elastic design to the current one of plastic design. The plastic design revolution has features that closely resemble a scientific revolution as described by Kuhn.
- Both a scientific theory and engineering model are representations of the world that are used to make predictions about it. However, because their goals (of *understanding* vs. *transformation*) and groundings (in *truth* vs. *safety*) are different, there are also significant differences in their bases, form, applicability, specification, methods of improvement and chief characteristics (*accuracy* vs. *dependability*).
- Kuhn's notions about the relativity of knowledge are not easily accepted by most practising scientists, because they believe largely in a world that is real, the truth about which can be found by inquiry, both theoretical and experimental. However, because engineers are more concerned about safety than truth, an engineering outlook can more easily accommodate the relativity of knowledge, provided that 'erring' from the 'target' is towards the safe side.

Acknowledgements Adapted from *The Structural Engineer 86*(2), 33–38, Paradigms, revolutions and models: some insights from Thomas Kuhn for an engineering outlook by W. P. S. Dias, 2008, published by the Institution of Structural Engineers, London.

References

W. Addis, *Structural Engineering: The Nature of Theory and Design* (Ellis Horwood, New York, 1990)

B. Barnes, *T.S. Kuhn and Social Science* (Columbia University Press, New York, 1982)

B. Barnes, D. Bloor, Relativism, rationalism and the sociology of knowledge, in *Rationality and Relativism*, ed. by M. Hollis, S. Lukes (MIT Press, Cambridge, 1982), pp. 21–47

D.I. Blockley, *The Nature of Structural Design and Safety* (Ellis Horwood, Chichester, 1980)

D.I. Blockley, Engineering from reflective practice. Res. Eng. Des. **4**, 13–22 (1992)

D.I. Blockley, Analysing uncertainties: towards comparing Bayesian and interval probabilities. Mech. Syst. Signal Process. **37**(1–2), 30–42 (2013)

J. Bobrowski, *Concrete Structures M.Sc. Lectures* (Imperial College, London, 1982)

W.P.S. Dias, Structural failures and design philosophy. Struct. Eng. **72**(2), 25–29 (1994)

W.P.S. Dias, Paradigms, revolutions and models: some insights from Thomas Kuhn for an engineering outlook. Struct. Eng. **86**(2), 33–38 (2008)

A. Einstein, Principles of research, *Mein Weltbild* (Querigo Verlag, Amsterdam, 1934), pp. 224–227

P. Feyerabend, How to defend society against science, in *Scientific Revolutions*, ed. by I. Hacking (Oxford University Press, Oxford, 1981), pp. 156–157

S.L. Goldman, Why we need a philosophy of engineering: a work in progress. Interdiscip. Sci. Rev. **29**(2), 163–176 (2004)

R. Hahn, *Anaximander and the Architects* (SUNY Press, Albany, 2001)

J. Heyman, *The Masonry Arch* (Ellis Horwood, Chichester, 1982)

P. Hoyningen-Huene, *Reconstructing Scientific Revolutions: Thomas S. Kuhn's Philosophy of Science* (University of Chicago Press, Chicago, 1993)

T.S. Kuhn, *The Structure of Scientific Revolutions*, 2nd edn. (University of Chicago Press, Chicago, 1970)

T.S. Kuhn, *The Essential Tension: Selected Studies in Scientific Tradition and Change* (University of Chicago Press, Chicago, 1977)

T.S. Kuhn, *The Copernican Revolution: Planetary Astronomy in the Development of Western Thought* (Harvard University Press, Cambridge, 1985)

C. Kulasuriya, W.P.S. Dias, M.T.P. Hettiarachchi, The aesthetics of proportion in structural form. Struct. Eng. **80**(14), 22–27 (2002)

I. Lakatos, History of science and its rational reconstructions, in *Scientific Revolutions*, ed. by I. Hacking (Oxford University Press, Oxford, 1981), pp. 107–127

D.A. Schon, *The Reflective Practitioner: How Professionals Think in Action* (Temple Smith, London, 1983)

S. Weinberg, The revolution that didn't happen. The New York Review, 8 Oct, pp. 48–52 (1998)

Chapter 5
Shared Values for Aesthetics and Ethics?

5.1 Can Values Be Measured?

We deal in this chapter with two areas that are considered to be essentially subjective, and also commonly called 'values'. Curiously, entities that are grouped under the term 'values' are precisely those that cannot be measured or given a number for. However, when we hold such values, whether they be of beauty (*aesthetics*) or morality (*ethics*), there is an expectation that others would share them too. In other words, we hold them with 'universal intent', in the words of Michael Polanyi (1891–1976). Another way to say this, after Heidegger, is that they form part of our shared experience (see Sect. 7.2). This chapter will draw extensively from the work of Polanyi, in particular from his major work *Personal Knowledge* (Polanyi 1958). This is because Polanyi wrote widely about the need for personal passion (to seek beauty) and moral responsibility (to arrive at truth that could be universally endorsed) in the pursuit of scientific discovery; as opposed to an attitude of skepticism and cold detachment.

We first apply to engineering some of Polanyi's insights on these relatively subjective areas of aesthetics and ethics. Then we argue that value judgments in these areas can be shared, and that they can therefore more properly be called 'inter-subjective' (see also Dias 2011). This will bestow importance on these aspects. Otherwise they run the risk of being downplayed, either because they cannot be measured; or because they are labelled as purely subjective and personal, without any voice in the 'public square'.

5.2 What Is an Elegant Solution?

Let us first deal with aesthetics, which engineers more easily recognize as *elegance*. The motto of the Society of Structural Engineers, Sri Lanka is "safety, elegance, economy". Safety and economy are readily recognized as being integral to engineering (Sect. 4.5), but so is elegance or aesthetics. Although visual beauty is very

P. Dias, *Philosophy for Engineering*, SpringerBriefs in Applied Sciences and Technology, https://doi.org/10.1007/978-981-15-1271-1_5

important in a discipline such as structural engineering, all forms of engineering look for elegance in their solutions, just as Polanyi argued that beauty was one of the major self-set standards of the scientist. For Polanyi (1958), 'knowing' was a problem solving activity that humans were driven to by a heuristic *passion*. This was the passion that for example directed the selection of a research problem. Animals too displayed a passion to solve a puzzle that confronted them, even when they did not receive a reward. One of the passions that drove human beings was a desire for *beauty*, which was reflected in many fields of human endeavour (Polanyi 1958, p. 133):

> The affirmation of a great scientific theory is in part an expression of delight. The theory has an inarticulate component acclaiming its beauty, and this is essential to the belief that the theory is true.... A scientific theory which calls attention to its own beauty, and partly relies on it for claiming to represent empirical reality, is akin to a work of art which calls attention to its own beauty as a token of artistic reality.... In teaching its own kinds of formal excellence science functions like art, religion, morality, law and other constituents of culture.

Scientists such as Kepler and Einstein had a great passion for intellectual beauty in their theories (Polanyi 1958). Feyerabend (1970) even advocated that theories be judged by their form—e.g. based on the number of steps required to derive a proposition from a theory—in addition to their content. Polanyi clearly understood that the practice of science was not value free—it was characterized by a search for beauty and excellence; and not merely for correspondence with nature. His argument was that the quest for the former would probably ensure the latter. Kuhn (1977) said that it was not possible to judge between competing paradigms using rationality alone, calling this the problem of 'incommensurability'. Some of the criteria he gave for making such judgements could be called 'rational'—e.g. accuracy, consistency and scope; but they also included simplicity (an aesthetic and metaphysical concept that resonates with Polanyi) and fruitfulness, which relates to (intuitive understanding regarding) the promise of continuing scientific activity.

Engineering is a problem solving discipline as we saw in Chap. 3, where it is sought to convert specifications to realizations. Such a realization, whether a product or system, is often judged with respect to its elegance. How then do we know that a solution is elegant? There are perhaps two aspects to elegance, associated with the two communities that would judge it. The first is the engineering community itself, whether as partners in a given project, or as observers. One of the chief characteristics of this elegance is *simplicity*, of which economy is a key ingredient, not so much in financial terms but in level of detail; *parsimony* may be a better word for it. Mathematicians in particular, and scientists too, actively espouse parsimony. A good example of parsimony from reinforced concrete design, is how all the reinforcement spacings in a water reservoir were made to be identical and equal to 170 mm; only the reinforcement sizes differed (Dias and Al-Kabbani 1997). Even though this may not have been the strictly least cost solution, it was certainly an elegant one, and would have made things very easy for the builders too. Another manifestation of parsimony from electronic circuit design would be to achieve the desired requirements with the minimum number of components and connections.

Although the above kind of elegance may be hidden from the public, they are also involved in judging engineering elegance. Here too simplicity may be important, but more so would be *functionality*. Consider the design of a mixing tap, which has to regulate both the flow of water as well as the relative proportions of the hot and cold streams. There are many realizations of this, some clearly more elegant than others. One important aspect of functionality is *intuitiveness*—the very design of the product must guide users towards its use. While this is merely useful for mixing taps, it is of paramount importance for process control systems. Another example of an engineering system is that of a subway station that serves multiple train lines. A good design will be one where passengers can easily find their way from one line to another (for the appropriate direction too) without getting lost! Where engineers are concerned, a particular subset of the public will be very special to them—namely their clients. Blockley and Godfrey (2000) speak about the importance of *delighting* the client; this can be regarded as the 'process' counterpart to beauty or elegance, which is primarily associated with a product or system.

Finally, in the 21st century world that prizes innovation highly, elegance could also refer to the success of the overall project, ranging from niche identification through conception and execution to market penetration. We sometimes call this a "neat solution". It demands an integration of wide ranging disciplines such as marketing, manufacturing, technology, design, management and law. Vojak et al. (2010) argue that the emergence of a successful innovation depends on a subsidiary awareness of the particular disciplines combined with a focal attention on the innovation. They draw a clear parallel between this and the tacit knowing advocated by Polanyi (1966) for scientific discovery, where scientists would 'dwell in' the particulars of various scientific theories in the process of focusing on a discovery they were seeking—see Sects. 5.5 and 8.2.

Engineering elegance arises therefore because it fits solutions to contexts, for which value judgements are required. According to Goldman (2017), such solutions are "constrained in various ways, to a degree by what nature will allow, but primarily by highly contingent factors that, from a logical as well as a natural perspective, are arbitrary: time, money, markets, vested interests and social, political and personal values" (see Sect. 1.1). Thus, engineering design solutions for a given context will differ from person to person, each of whom will argue that their design is the 'best'. However, broad consensus can nevertheless be held regarding excellence in such designs, which is why they are often submitted for competitions conducted by professional institutions.

5.3 Differing Views of Aesthetics

Although aesthetic values can be shared, we could expect to find that such values differ from one community to another. We could also expect the notion of beauty or elegance to change with time. One of the philosophical positions held regarding modern technology is that it is different from an earlier pre-modern *techne*, the

Greek word for art, skill or craft. According to Heidegger (1977), there was no difference at that time between craftsmanship and poetry, primarily because there was intimate human involvement in both. Also, because the tools themselves and their transforming capacity were limited, the products of such technology were never too far removed from the earth (see Sect. 6.3). This generated a particular type of aesthetic, which is valued even today. For example, a good violin needs to be hand crafted by a master craftsman.

By and large however, mass production and widespread mechanization in the modern world have alienated humans from both their tools and products, leading to alienation from the earth. Our tools are far more 'efficient' than we are, leading to a crisis in human worth, and to labor being treated as a mere commodity in the process of production. Our products are also highly processed from their original states in the earth. Modernity and mechanization have therefore been accused of destroying aesthetics for the sake of efficiency. This can be considered as form being subjugated to function. We shall return to Heidegger's contribution to this debate in Chap. 6.

There is a view however that ultra-modern technology has reversed this trend and created aesthetic values of its own. The reason for this is a separation of form from function. As one writer has put it, 'style follows sales' rather than 'form following function' (Rutsky 1999). Two of the styles he identifies are minimalism on the one hand and complexity on the other. Minimalism has perhaps arisen from the widespread use of plastics, where the easy moulding of shapes can reduce many of the edges and joints in say household products such as chairs. Complexity can be seen for example in the vast array of hierarchically organized options and drop down menus now available in computer software and operating systems. Miniaturization can be identified as another ultra-modern style, especially in the micro-electronics industry.

Modernity can also be contrasted with the post-modern world of today. When aesthetics are based on 'theoretical frameworks', we can say that the aesthetics reflect modernity (see also Sect. 6.2). The forms of the artefacts or structures that arise from such aesthetics will reflect their function. The Millau viaduct over the River Tarn in France, the tallest bridge in the world, had to be designed according to the exacting demands of engineering theory, but has resulted in a most graceful cable stayed bridge. Post-modernism on the other hand can be seen as a movement that rejects 'theory', resulting in a diversity of styles that probably 'follow sales'. Frank Gehry's Guggenheim Museum in Bilbao is a 'free form' structure that gives no indication whatsoever of the load paths in the structure. Nevertheless, its iconoclastic style has an aesthetic of its own, and has reportedly resulted in an economic revival of the entire region.

Apart from the above, there are also differing models of aesthetics that reflect different schools of thought on what constitutes beauty. Kulasuriya et al. (2002) describe some of them, two of which are:

(i) *Objective Idealism*: According to this model, aesthetics is an objective quality, which is independent of and external to human consciousness. It is often associated with concepts that are considered 'universal', for example the notion that a circle or square is 'perfect' because they reflect the concept of equality.
(ii) *Metaphysical Naturalism*: This model regards objects to be aesthetic inasmuch as they reflect properties—whether of proportion, composition, shape or even colour—that are found in nature.

5.4 An Example of the Aesthetics of Proportion

We shall now look at an example where form follows function in an aesthetically appropriate way. Consider a bridge with three spans. In general, some sort of symmetry would be required to make the bridge aesthetically appealing, so we can set the two end spans to be equal in length. But which ratio of central to end spans would be 'most pleasing' to the eye? Opinions on this were obtained from a set of 50 civil engineers—details are given by Kulasuriya et al. (2002). The intention was to explore whether the choices made by the engineers correspond to (i) ratios derived from models of aesthetics; and (ii) economically suitable ratios. Figure 5.1 shows a frequency plot of the ratios chosen by the respondents. The results show that the most preferred ratio is unity—i.e. equal span continuous bridges. There is however another maximum at a ratio of 1.6. This is very close to the value of $(1+\sqrt{5})/2 = 1.618$, referred to by the ancient Greeks as the 'golden mean'. It is found when a line is divided into two unequal lengths so that the shorter relates to the longer as the longer relates to the whole. The golden mean proportion appears frequently in nature, from sunflowers, apple blossoms, and pinecones in the plant world to spiral shells beneath the seas. Many of the proportions of the human body also conform to the golden mean.

The desire for the golden ratio may be linked to an 'interest in nature', and corresponds to the aesthetic model of *metaphysical naturalism*, which regards aesthetic

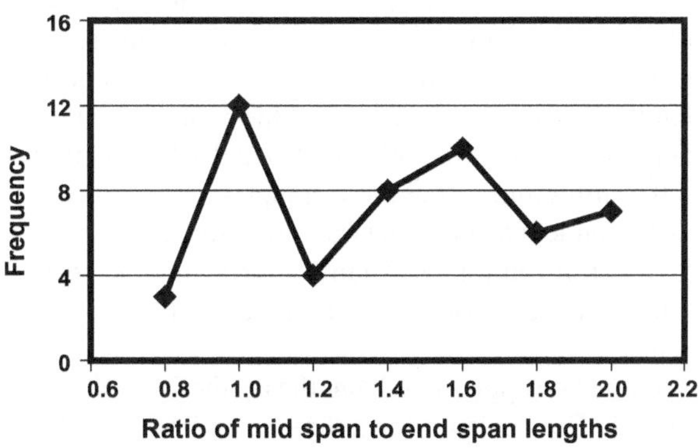

Fig. 5.1 Survey respondents' aesthetic preferences for span ratios in 3 span bridge (from Dias 2011)

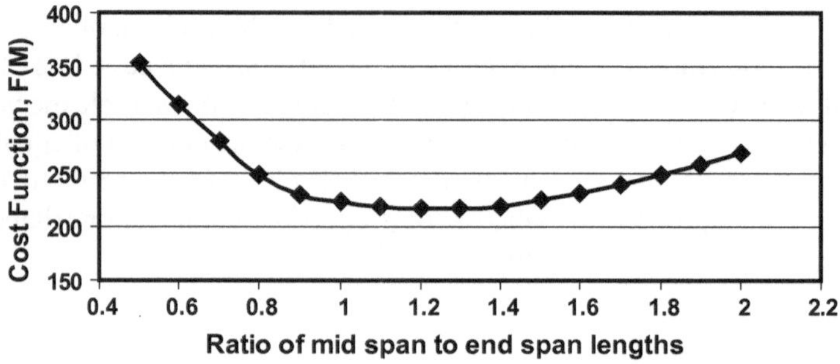

Fig. 5.2 Cost variation with span ratio in 3 span bridge (from Dias 2011)

norms as a reflection of nature. The desire for a ratio of unity might have addressed the respondents' sense of perfection, and corresponds to the aesthetic model of *objective idealism*, which regards aesthetic proportions as reflecting 'ideal' shapes or forms. The greater preference for the equal span bridge emphasises the power of the concept of equality. This preference was shown despite the fact that the respondents were civil engineers whose training would have suggested that shorter end spans would result in a more economical distribution of bending moments and stresses. Although they were asked not to consider economy, their subconscious thoughts may have been expected to affect their preferences.

Figure 5.2 shows the cost curve, for various ratios of mid span to end span lengths. It was assumed that (i) in end spans the sagging moments are effective over a length of 0.8 times the end span from exterior supports; (ii) in the mid span the sagging moment is effective over a length of 0.6 times the span at the middle; and (iii) at interior supports hogging moments are effective over a length of 0.2 times the span on each side of the support. (Note that sagging moments cause a convex downward curvature and hogging moments an upward one.) The simplified cost function based on the above shows a minimum at a proportion around 1.3, a value close to $\sqrt{2}$ (another incommensurable ratio considered to be 'aesthetic' by the ancient Greeks), and roughly equidistant from both unity and the golden ratio.

So, although not all respondents chose ratios that were based on aesthetic theory, the two peaks in the frequency distribution did correspond to ratios of aesthetic significance based on models of aesthetics. This shows that there are largely accepted norms even in the relatively subjective area of aesthetics. The exercise also showed that the economic minimum (at the ratio of 1.3) was equidistant from the two 'aesthetic peaks' (i.e. the ratios of 1.0 and 1.6); and that the cost function was also close to the minimum in the entire range between these ratios. This shows a remarkable correspondence between structural efficiency on the one hand and aesthetic form on the other. It reinforces Polanyi's ideas that the search for beauty in the human mind will often correspond to truth in nature. Engineers would say, "if it looks right, it *is* right".

It must be appreciated that the aesthetic choices in the above example were highly limited. The decision for a three span bridge had already been made. A client who wanted an iconic structure may have preferred a suspension bridge; a river with one half of its cross section having a deep river bed may have required a two span bridge. These constraints would have called for different solutions, emphasising again the contingent nature of engineering design, and the elegance required for addressing it. However, within a set of limited choices, we were able to see shared aesthetic values.

5.5 Morality and Faith

According to Polanyi, there were three spheres in which morality was to be displayed in the practice of science. First, individual scientists had to exercise morality in 'making sense' of their data, since the knowledge they were discovering had universal intent. In other words, scientists performed the 'integrations' of their clues to arrive at conclusions regarding an external reality they felt others could arrive at too. This required a commitment equivalent to a personal *moral* responsibility, which a mechanical process of scientific induction would not require. Polanyi (1958, pp. 308–9) likened this to the decision arrived at by a judge:

> We can watch the mechanism of commitment operating on a minor scale, and yet revealing all its characteristic features, in the way a judge decides a novel case. His discretion extends over the possible alternatives left open to him by the existing explicit framework of the law, and within this area he must exercise his personal judgment. But the law does not admit that it fails to cover any conceivable case. By seeking the right decision the judge must find the law, supposed to be existing – though as yet unknown. This is why eventually his decision becomes binding as law. The judge's discretion is thus narrowed down to zero by the stranglehold of his own universal intent – by the power of his responsibility over himself.

We see here that the freedom of the *subjective* person to do as he pleases is overruled by the freedom of the *responsible* person to act as he must. Note however, that there would have been no responsibility if there was no independence. Another way of putting it is that scientific progress required a scientific *conscience* in addition to creative impulses and critical caution (Polanyi 1946).

Secondly, morality was required within the scientific community in the process of peer-review. This can be called the communal sphere of morality, where the community fulfilled an *authoritative* role (in addition to, or as part of maintaining a tradition). Polanyi (1946) took great pains to demonstrate that scientific authority was distributed as opposed to concentrated. There was no pyramid of authority in science, primarily because no scientist could claim to be competent in more than a tiny fraction of scientific knowledge. Science was like a chain of overlapping beliefs, where knowledge for a scientist, other than in his specialized field, was second hand at best (Polanyi 1958).

The authority of science was displayed not so much in initiating or *directing* various activities, but rather by *denying* opportunity for certain avenues to be explored. This took place for instance, by the turning down of an application for a research grant or the rejection of a manuscript for publication, after a process of peer-review (Polanyi 1946). Polanyi (1958) said that scientists judged each other on trade-offs between considerations such as validity (or accuracy), systematic relevance (or profundity) and intrinsic interest. However, such peer-review still required the exercise of personal judgment, informed by moral conviction—to ensure, for example that a competitor's application was not unfairly marked down.

Thirdly, the wider society too had to be supportive of the scientific enterprise, believing that there was such a thing as truth and that it could be discovered by scientists (Polanyi 1946). This can be called the social sphere of morality. Scientific research is supported largely by public funds—for even very fundamental research; but that means it cannot be used for other pressing social needs. Public support was essential therefore for science to flourish.

Polanyi also wrote about the importance of 'faith', a concept normally associated with religious belief, just as morality is (Dias 2010). One of his books is titled *Science, Faith and Society* (Polanyi 1946). He used the word 'fiduciary' extensively for describing how science should be practiced; and also said that all experience, including that of scientists, was based on belief. He likened this belief to religious faith, and quoted St. Augustine who said "*nisi credideritis, non intelligetis*" (if you do not believe, you shall not understand) (Polanyi 1958, p. 266). What this means is that we need to 'dwell in' our theories to use them for making discoveries. Polanyi (1966) called this process 'tacit knowing'. He likened this to a blind man who 'dwelt in' his stick (or vice versa since the stick became an extension of his arm) in order to explore a cave. However, the blind man was aware of the limitations of his stick—e.g. if and when the dimensions of the cave exceeded the reach of his stick. So also we were to believe in our theories only provisionally. Apart from the practice of dwelling in our theories, the entire enterprise of science itself can only be carried out because we *believe* in certain assumptions (Ramachandra 2008, p. 181):

> …namely (1) that there is a real world outside our minds, and that the world is structured in an orderly and intelligible way; (2) that this rational order is contingent, it cannot be deduced in advance by logical reasoning but has to be discovered, thus calling for a basic posture of humility before the world whose rationality we seek to articulate through our theories and experiments; and (3) that the intelligibility of the universe is accessible to the human mind: our epistemic abilities, though not unlimited, are adequate to this task.

One of the roles of the scientific community was to transmit such faith to the next generation of scientists, presumably in the form of the 'ruling paradigm' in Kuhn's words—note Kuhn's comparison of science to orthodox theology (Sect. 4.2). In addition, this community was made up of 'master craftsmen' in their disciplines, under whom budding scientists could receive 'apprenticeship'. Much scientific discovery now takes place while young scientists train under their professors to obtain doctoral degrees. In many ways the methodologies of research cannot be taught (although such courses abound in the academy), but rather 'caught' from a practitioner.

5.6 Engineering Ethics

In technology and engineering, the term *ethics* is used more commonly than that of morality. Ethics can have differing foundations or be based on different theories, just as there are different models of aesthetics. Three main ethical theories have been identified, namely utilitarian, deontological and virtue ethics (Blockley 2005). Utilitarianism is a viewpoint that judges an action based on its consequences, often with the aim of maximizing happiness for the greatest number; and most public policy is probably based on some notion of utilitarianism. There is however the issue raised by the philosopher Karl Popper (who we encountered in Chap. 3) that we should call for 'minimizing unhappiness' rather than the somewhat utopian goal of 'maximizing happiness' (Magee 1973). Popper (1945, pp. 284–5) said:

> I believe that there is, from the ethical point of view, no symmetry between suffering and happiness, or between pain and pleasure.... Human suffering makes a direct moral appeal for help, while there is no similar call to increase the happiness of a man who is doing well anyway...Instead of the greatest happiness for the greatest number, one should demand, more modestly, the least amount of avoidable suffering for all.

Deontological ethics is based on certain norms and maxims—e.g. "honesty is the best policy"; "avoid all conflicts of interest" etc.—that often arise from a person's upbringing and/or religious beliefs. Virtue ethics (to which pragmatism is closely related) is more context dependent and constitutes actions that make a person more virtuous—e.g. risking one's reputation for integrity by working in a corrupt environment so as to bring change. The complexities of engineering decision making may require a virtue ethics approach that recognizes shades of gray between extreme black and white ethical poles.

Engineering ethics concerns have also been divided into micro and macro issues. Micro issues involve the behaviour of single individuals, such as responsibilities towards the client, the public and fellow professionals. These are largely internal to the profession, and involve aspects such as competence, trustworthiness, honesty and fairness. Macro issues encompass the impact of engineering or technology on a wider scale. These are treated as external to the profession, and focus mainly on the protection of public safety, health and welfare (Herkert 2001). One issue that links both the micro and macro spheres is the conflict of interests—i.e. when a decision that is supposed to be taken in the public interest by an individual (in the capacity of an appointed official) could also result in a direct benefit to that individual (Wells et al. 1986). For example, engineers who work for the state's approving or regulating agencies should not receive any pecuniary benefit from agencies whose proposals or work they approve or regulate.

Engineers have not been so clear on some of the subtler macro issues, and have hence been critiqued by social analysts for the unintended effects of technology. Some issues that can be identified are the impact of technological development on the environment, the influence of automation and robotics on human psychology, the effect of information technology on human community (including issues of privacy), the risks of nuclear technology whether in war or peace, the threat to human dignity from genetic engineering, and the role of technology in wealth creation and distribution (Ferre 1995). Many of these issues have been brought up, not by engineering bodies, but by wider social groups seeking to curb technology, or at least its ill effects. In many cases such reactions are too late, because the technology is too far developed, as for example in nuclear technology. In other cases, such as in environmental activism, the antagonism between environmentalists and technologists is implacable to the extent that it is difficult to work towards a 'common good'.

There is a strong case for educating engineers themselves to see such macro ethics as part of their own concerns. This will enable them to take part in debates regarding how their artefacts best serve the public, including questions of where investment in physical infrastructure is best directed. Otherwise, most engineers will view their work as value free, and focus on the technical challenges of their profession. However, because their activity will be reflected in a social context, often far more than that of pure scientists, ethical or moral choices should be exercised at the individual level itself. This resonates with the notion that engineers seek safety while scientists seek truth (Sect. 4.5).

One of the problems in implementing such an ethic may be that it will call, in some cases, for an ethic of *inaction*—something that may go against the very grain of a pragmatic and creative engineer (I. Nair, personal communication, 2001). There are however, many calls for such taboos, even from within the ranks of technologists themselves, on technologies such as robotics, genetic engineering and nanotechnology (Joy 2000); and more recently on some applications of artificial intelligence (Lawrence et al. 2016). Such calls may work if ethically educated and sensitive engineers pursue alternative technologies from *within*, rather than being imposed upon from *without*. We shall return to these imperatives in Sect. 6.4, where it is suggested that engineers should 'question in practice' as part of their 'reflection in practice'. We also present engineering ethics in Sect. 6.5 as an ethic of care (Pantazidou and Nair 1999).

5.7 The Engineer, the Public and the Professional Institution

Just as Polanyi identified three spheres of morality for science, engineering ethics often involves an interplay among three entities, namely the individual engineer, the professional institution and the general public. Although we have spoken earlier in this chapter about delighting the client (who would be a member or members of the public), for many *individual* engineers their most direct client is their employer. Since this corporate client or organization is often the engineer's sole employer, the pressure to look after the employer's interest is great. This is compounded by the fact that engineers tend, more than most other professionals, to work together in large numbers in a single organization, where there would undoubtedly be competition for promotion as well.

On the other hand, it is the public that uses the engineer's products, whether buildings, automobiles or appliances. This general public also serves as a forum for judging technology. Probably no other profession has such an influence on the everyday lives of people; conversely however, engineers do not interact directly with the public in the way that doctors, lawyers or even architects do. This 'facelessness' has caused an under-appreciation of the engineer's role in our society. At the same time, the importance of their responsibility to the public is not be felt by them in a direct way. The issue is complicated by the possibility that there could be conflicts of interest between the client and the public. In particular, this conflict could be manifested as the tension between economy (client's interest) and safety (the public interest).

Two examples of disasters will serve to highlight the pressures that engineers are under to please their (employer) clients. The first was the Challenger space shuttle disaster, where an engineering manager, who wanted to postpone the launch because of safety concerns, subsequently changed his mind when asked to "think like a manager"—in other words, to look after his employer's interests—with a 'calculated risk' (Pinkus et al. 1997). The second was the market release of the economical and fuel-efficient Ford Pinto motor car, despite the engineers knowing that there was a lack of integrity in its fuel system that could cause it to go up in flames if the car was 'rear-ended' in an accident (Birsch and Fielder 1994). Although the car was a 'best seller', it was involved in many deaths and injuries. It transpired later that the company concerned had considered it cheaper to pay the anticipated compensations rather than to incur expenditure for improving the design—although it subsequently had to pay out much more than the cost of improvement.

Professional institutions are supposed to play the role of regulating the practice of engineers by ensuring that the public interest is looked after—largely through their codes of conduct. Most such institutions all over the world place the engineer's responsibility to the public above that of their obligations to clients (Herkert 2001). This consensus could be seen as demonstrating the inter-subjective agreement among engineers where ethical values are concerned. The engineers' membership in such institutions, the ethical codes of which their fellow engineers and even employers (if engineers) would abide by too, could give them the confidence to defy their employer clients in the public interest if the need arises (Davis 1998). In practice however, the evidence indicates that professional institutions are unwilling or unable to support individual engineers in this area (Herkert 2001). Hence such individuals are forced to choose between compromising their standards or fulfilling their ethical obligations by 'whistle-blowing' (Oliver 2003).

Finally, just as *faith* is indispensable for the practice of science, the practice of engineering requires *trust*, because engineering is not only about products, but also involves purpose and people. All of this can be seen as incorporated in the engineering process. The engagement with people permeates all of engineering. It influences purpose, because engineers create products for use by people; and at times even misuse or abuse by them, whether accidental or deliberate (Blockley and Dias 2010). Process is also intimately linked to people, whether in jointly arriving at purpose or working together to deliver the product. Such interpersonal relationships are impossible without trust. As Blockley and Dias (2010) put it: "Ethics is also about our relationships with others—it arises from social contact with others. We learn, we reflect, we influence others and hopefully we develop that most precious, but most fragile, characteristic of good relationships which is trust". Where both faith and trust are concerned, certainty is impossible; but it is risk taking in the face of such uncertainty that delivers success.

5.8 Summary

- Michael Polanyi's ideas regarding the aesthetics of scientific theories and the three spheres of scientific morality have parallels in engineering elegance and ethics respectively.
- Engineering elegance arises from the tailoring of solutions, characterized by simplicity and functionality, to suit a variety of contexts. Such contexts will in turn necessitate a variety of value based decisions relating to time, money, vested interests and socio-political pressures.
- Nevertheless, where aesthetics is concerned, an example was presented of shared consensus regarding the visual appearance of a bridge; a consensus that was consistent with structural efficiency and economy too.
- It was possible to map Polanyi's spheres of morality involving the individual scientist, scientific community and wider society directly to the ethical imperatives associated with the engineering practitioner, professional institution and general public. In many ways the wider society is much more directly impacted by engineering than by science; and the major stated concern of most engineering institutions worldwide is the safety of the public.
- A parallel can also be found between Polanyi's advocacy of *faith* as a requirement for scientific practice and the hallmark of *trust* that is essential for the practice of engineering.
- Where both aesthetics and ethics are concerned, examples were given for a reasonable degree of agreement among practitioners, thus encouraging their exploration as inter-subjective disciplines, as opposed to labelling them as totally subjective and thus devoid of significance for public life.

Acknowledgements Adapted by permission from Springer Nature Customer Service Centre GmbH: Springer Nature: *Science and Engineering Ethics, 17*(2), 233–243: Aesthetics and ethics in engineering: insights from Polanyi, by Priyan Dias, 2011.

References

D. Birsch, J.H. Fielder (eds.), *The Ford Pinto Case: A Study in Applied Ethics, Business and Technology* (SUNY Press, Albany, 1994)

D. Blockley, Do ethics matter? Struct. Eng. **83**(7), 27–31 (2005)

D. Blockley, P. Dias, Managing conflict through ethics. Civ. Eng. Environ. Syst. **27**(3), 255–262 (2010)

D.I. Blockley, P. Godfrey, *Doing It Differently: Systems for Rethinking Construction* (Thomas Telford, London, 2000)

M. Davis, *Thinking Like An Engineer: Studies in the Ethics of a Profession* (Oxford University Press, New York, 1998)

P. Dias, Is science very different from religion? A Polanyian perspective. Sci. Christ. Belief **22**(1), 43–55 (2010)

P. Dias, Aesthetics and ethics in engineering: insights from Polanyi. Sci. Eng. Ethics **17**(2), 233–243 (2011)

W.P.S. Dias, I. Al-Kabbani, Design and performance of 11350 m^3 rectangular Jubilee Reservoir in Sri Lanka. Engineer, Sri Lanka, (June), pp. 74–81 (1997)

F. Ferre, *Philosophy of Technology* (University of Georgia Press, Athens, 1995)

P. Feyerabend, Consolations for the specialist, in *Criticism and the Growth of Knowledge*, ed. by I. Lakatos, A. Musgrave (Cambridge University Press, Cambridge, 1970), pp. 197–230

S.L. Goldman, Compromised exactness and the rationality of engineering (Chap. 1), in *Social Systems Engineering; The Design of Complexity*, ed. by C. Garcia-Diaz, C. Olaya (Wiley, Oxford, 2017), pp. 13–29

M. Heidegger, *The Question Concerning Technology and Other Essays* (trans: W. Lovitt) (Harper and Row, New York, 1977)

J.R. Herkert, Future directions in engineering ethics research: Microethics, macroethics and the role of professional societies. Sci. Eng. Ethics **7**(3), 403–414 (2001)

B. Joy, Why the future doesn't need us. Wired, Ideas 04.01.00 (2000), https://www.wired.com/2000/04/joy-2/. Accessed 12 Dec 2018

T.S. Kuhn, *The Essential Tension: Selected Studies in Scientific Tradition and Change* (University of Chicago Press, Chicago, 1977)

C. Kulasuriya, W.P.S. Dias, M.T.P. Hettiarachchi, The aesthetics of proportion in structural form. Struct. Eng. **80**(14), 22–27 (2002)

D.R. Lawrence, C. Palacios-Gonzales, J. Harris, Artificial intelligence: the Shylock syndrome. Camb. Q. Healthc. Ethics **25**, 250–261 (2016)

B. Magee, *Popper* (Fontana, London, 1973)

D. Oliver, Whistle-blowing engineer. ASCE J. Prof. Issues Eng. Educ. Pract. **129**(4), 246–256 (2003)

M. Pantazidou, I. Nair, Ethic of care: guiding principles for engineering teaching and practice. J. Eng. Educ. 205–212 (1999)

R.L.B. Pinkus, L.J. Shuman, N.P. Hummon, H. Wolfe, *Engineering Ethics: Balancing Cost, Schedule and Risk—Lessons Learned from the Space Shuttle* (Cambridge University Press, Cambridge, 1997)

M. Polanyi, *Science, Faith and Society* (University of Chicago Press, Chicago, 1946)

M. Polanyi, *Personal Knowledge: Towards a Post-critical Philosophy* (University of Chicago Press, Chicago, 1958)

M. Polanyi, *The Tacit Dimension* (Doubleday & Co., Garden City, 1966)

K.R. Popper, *The Open Society and Its Enemies*, vol. 1 (Routledge & Kegan Paul, London, 1945)

V. Ramachandra, *Subverting Global Myths: Theology and Public Issues Shaping Our World* (IVP Academic, Downers Grove, 2008)

R.L. Rutsky, *High Techne: Art and Technology from the Machine Aesthetic to the Posthuman* (University of Minnesota Press, Minneapolis, 1999)

B.A. Vojak, R.L. Price, A. Griffin, Corporate innovation, in *Oxford Handbook of Interdisciplinarity*, ed. by R. Frodeman, J. Klein, C. Mitcham (Oxford University Press, Oxford, 2010), pp. 546–559

P. Wells, H. Jones, M. Davis, *Conflicts of Interest in Engineering* (Kendall/Hunt Publishing, Dubuque, 1986)

Chapter 6
Is Technology Neutral?

6.1 Suspicion and Questioning

We saw in Sect. 5.5 that Michael Polanyi promoted an attitude of *faith*, without which it would not be possible to hold scientific theories or sustain a scientific community. Martin Heidegger (1889–1976) on the other hand advocated a spirit of *suspicion* towards our evaluation of the world and its workings. This is because our routine approach to life, which he called 'average everydayness' (Heidegger 1997), often *concealed* the true nature of things, though at times *disclosing* them. The practical way in which he exercised suspicion was to raise questions about things we normally took for granted. In that sense he can be considered an intellectual descendent of Socrates himself, who was called the 'gadfly of society' because of his incessant questioning.

Heidegger's early writings stressed the primacy of practice over theory, and hence can be interpreted as supportive of the engineering approach, as will be argued in Chap. 7. His later writings questioned the foundations of modern technology, as explained in this chapter (see also Dias 2003). As we shall see, the term 'technology' encompasses not only engineers (who are its main agents), but also an entire network of relationships and worldviews that today are called rationalistic or mechanistic or digital in various contexts. One of Heidegger's (1977) collection of works is titled *The Question Concerning Technology and Other Essays*. He was also preoccupied with poetry, which he considered to be an antidote to modern technology—another of his collections is called *Poetry, Language, Thought* (1971). Polanyi and Heidegger proposed very different outlooks to things. However, there is a curious similarity in their views of what can be called 'practice based knowledge', as we shall see in Chap. 8.

P. Dias, *Philosophy for Engineering*, SpringerBriefs in Applied
Sciences and Technology, https://doi.org/10.1007/978-981-15-1271-1_6

6.2 Technology and 'Enframing'

Heidegger's (1977) essay on *The Question Concerning Technology* deals directly with the issue as to whether technology is neutral. We mean by 'neutrality' that technology is neither good nor bad; or that it becomes so depending only on the way it is used by humans. Heidegger however, cautioned against thinking that things (like technology) were 'neutral'. He also said that we should go beyond what was 'merely correct' to seek what was 'true'. The essay is a good example of his spirit of suspicion and commitment to questioning. The word 'hermeneutics' is used to describe the process of interpretation, whether of texts or events; and Heidegger is said to have had a 'hermeneutic of suspicion' (Dreyfus 1991)—in other words he interpreted the world to himself through a lens of suspicion.

Mitcham (1994) has made one of the best philosophical expositions on the nature of technology, where he refers extensively to Heidegger. For Heidegger, technology had *instrumental* and *anthropological* aspects; meaning that it was both a means to an end and a human activity respectively. In addition, using the artefact of a silver chalice as an example, Heidegger (1977) argued that technology was a way of *revealing*. The instrumental (or means-to-an-end) aspect of technology was based on causality; and Heidegger presented Aristotle's four causes for an artefact—namely the material, form, purpose and agent. The purpose is often what we think of first—if we want an object from which to pour out sacrificial liquid offerings, we need a cup-like device. But because it is part of a ceremonial procedure, we will probably use silver, a precious metal, to make it; and give it an impressive looking form or shape. However, in order for the chalice to become a reality we need the agency of a silversmith as well.

The idea of causality is derived from the Latin word *cadere*, which meant 'to fall out as a result' (Heidegger 1977). The human activity in technology served to gather together the causes, all of which were quite different in nature and character. The material is physical; the form conceptual; the purpose mental; and the agency human. This 'gathering' of rich variety and 'falling out' constituted a 'bringing forth' (*poiesis* in Greek, which is the root of the word 'poetic') from concealment into 'unconcealment' (*aletheia* in Greek, which also meant 'truth'); and was hence a *revealing*. According to Heidegger (1977), the Greek word for technology, *techne*, was therefore correctly used for fine arts such as poetry as well, which was a bringing forth of a different kind. He also said that until Plato's time, *techne* was linked with *episteme* (Greek for 'knowledge'). We see here a link not only between technology and poetry, but also between such 'bringing forth' and knowledge and truth. Truth and knowledge have a very practical connotation here, as opposed to a theoretical one. It can be called 'knowing *how*' as opposed to 'knowing *what*', and we explore this in Chap. 7.

Heidegger (1977) made a distinction however between *techne* and modern technology. While both were ways of revealing, he said that modern technology was not a 'bringing forth', but rather a *challenging*. This was primarily because modern technology 'levelled down' everything, including humans, to the status of a *resource*, or '*standing-reserve*' as he put it. So how did an outlook that produced *techne* change to one that created modern technology? Perhaps it was due to the Greek idea of 'work' (which involved a rich human involvement in 'disclosing' an artefact) being translated as *fungere* in Latin, from which we get the idea of 'function' (which is an abstraction from physical reality)—i.e. many realizations can perform the same function, and all such realizations are 'labelled' with the same function.

Thus 'The Rhine' was no longer a river for writing poems about, or even the river that was crossed by an old wooden bridge; rather, it became labelled as an 'input' to a hydroelectric power station. Similarly, the forester in the woods became an input to the cellulose production process. Treating everything as 'resources' or inputs means that one can be exchanged for another, with little or no consideration for the disruptions caused by such changes. So if the forester is replaced by a harvesting machine, the former loses his means of livelihood and is also divorced from his close-to-the-earth work environment. In addition, the machine may harvest the forest at a rate that eventually destroys it, thus harming both future cellulose production from that region and also everything else the forest stood for—ranging from carbon capture through biodiversity preservation to human recreation.

Today the word 'work' itself is used as a label for what can be done with energy—we could say that energy is the 'capacity to do work'—whatever the source of energy or type of work. This 'levelling down' to a common entity is called 'reductionism'—an example of this is to describe a glorious sunset as 'nothing but' a collage of light rays with different wavelengths. Note the difference between such reductionism and the rich variety of causes contributing to the silver chalice. Heidegger identified the seeds of this reductionism in mathematical physics, which though pre-dating modern technology, displayed the tendency to level things down by abstracting from the world only entities that could be represented mathematically—see the quotation about theoretical physics by Einstein in Sect. 4.5, where he insists on (mathematical) 'simplicity' in scientific theories. We have argued there that engineering models need rather to capture 'completeness'—very much in line with Heidegger. Reductionism and abstraction have made huge contributions to science based technological advancement. However, Heidegger was pointing out its negative consequences.

Humans are entrapped in this process of challenging nature, because it also includes themselves; this was called '*enframing*' by Heidegger (1977). Such enframing posed many dangers. On the one hand, man also became 'standing-reserve', while at the same time he exalted himself to being the 'lord of the earth'—since it was man himself who performed the abstraction. Furthermore, although modern technology in the form of enframing was also a revealing of being, its pervasiveness threatened to erase all other modes of revealing. Zimmerman (1990) has suggested that Heidegger could be called a 'deep ecologist' because of his attempts to erase subject-object distinctions; his concern for 'gathering' entities having different qualities rather than reducing everything down to a 'resource' or 'input' (see above); and his concept of 'care' as the human stance towards the world, in the sense of taking care of things—all of these aspects are important for philosophically serious (or 'deep') ecologists. Some of these ideas are tackled below, but others once again in Chap. 7.

At any rate, Heidegger advocated that modern technology should be confronted from a realm that was similar to it, yet fundamentally different. He thought that art was such a realm—different to technology, yet similar in the sense of 'bringing forth'—and much of his later writing was devoted to this aspect. However, he considered that the mere act of questioning had a salutary effect too (Heidegger 1977); and we return to this in Sect. 6.4.

6.3 Art and Poetry

Heidegger's best known reflection on art is that concerning a pair of shoes, supposedly of a peasant woman, in a Van Gogh painting. He said that while a work of art such as this did possess the character of a 'mere thing' as well, it was primarily a revealing. This revealing took place because the peasant's shoes constituted a 'setting forth of the earth' as well as a 'setting up of a world'. In other words, we were able to enter the *world* of the peasant, which was rooted in an *earthy* context. So the work of art brought together these elements and 'being' was revealed through such *gathering*. Art was therefore much more than an imitation or depiction of reality (Heidegger 1971). Consider this description, which illustrates the revealing that takes place through a work of art (Heidegger 1971, p. 33):

> From the dark opening of the worn insides of the shoes the toilsome tread of the worker stares forth. In the stiffly rugged heaviness of the shoes there is the accumulated tenacity of her slow trudge through the far-spreading and ever-uniform furrows of the field swept by a raw wind. On the leather lie the dampness and richness of the soil. Under the soles slides the loneliness of the field-path as evening falls. In the shoes vibrates the silent call of the earth, its quiet gift of the ripening grain and its unexplained self-refusal in the fallow desolation of the wintry field. The equipment is pervaded by uncomplaining anxiety as to the certainty of bread, the wordless joy of having once more withstood want, the trembling before the impending childbed and shivering at the surrounding menace of death. This equipment belongs to the *earth*, and it is protected in the *world* of the peasant woman. (Italics from original)

The inadequacy or poverty of a 'scientific' description is clear when compared to this. This is probably because science is analytical, stripping away outer layers of context and life, while art is a gathering. Furthermore, most art, including painting and poetry, both of which Heidegger was preoccupied with, is metaphorical rather than literal; and can accommodate many interpretations unlike scientific descriptions. In Sect. 6.5 we consider how metaphor can be used to generate creativity in design.

The reference to 'earth' (and hence nature) is also a critique of any modern technology that seeks production in unnatural ways—for example intensive poultry farming, where chickens are merely 'meat machines'; or even simply distances us from contact with the earth—e.g. in artificial air-conditioned environments where all transactions occur in cyberspace. Furthermore, because gathering is a revealing or unconcealment (*aletheia* or 'truth'), and because art in turn is associated with beauty, Heidegger said that "beauty is one of the ways in which truth occurs as unconcealedness" (Heidegger 1971, p. 56). In his description of a wine jug, Heidegger described this gathering as one of 'earth and sky', 'mortals and divinities'—'earth and sky' because of their contributions to both jug and wine, and 'divinities and mortals' for their participation in libation and drink respectively. This 'quaternity' was identified in a temple, bridge and building too (Heidegger 1971).

6.4 Questioning in Engineering Practice

Given the pervasive and significant impact of technology on our lives and society, engineers too should engage in such questioning as an integral part of their practice, since they are agents of technology. This would also result in more balanced critiques of technology. Today, critics of technology are largely philosophers or environmentalist, both of whom are sometimes unrealistic in their rejection of technology. Feenberg (1999), though himself a philosopher, critiques Heidegger for an 'essentialist' view of technology—i.e. for holding that the essence of technology is a reduction to function and inputs. He says rather that we should accept the fact that we are part of technology, and make appropriate interventions to defend a meaningful life and a livable environment. While philosophers like Heidegger seem to want a *rejection* of technology in favor of a return to nature, or at least to a 'gentler' *techne*, modern environmentalists actually express a *reliance* on modern technology—e.g. for cleaner burning of coal. The main message of the environmentalists is that technology is not always *benign*. The deeper message of the philosophers is that technology is not *neutral*. Heidegger (1977, p. 4) said for example:

> But we are delivered over to it [i.e. technology] in the worst possible way when we regard it as something *neutral*; for this conception of it, to which today we particularly like to do homage, makes us utterly blind to the essence of technology. (Parentheses and italics by present author)

We shall consider four levels at which technology influences us, going from the shallower and more visible to the deeper and less obvious. The first two demonstrate the concerns of environmentalists (and others concerned with justice and politics) and the latter two those of philosophers (and social analysts). The first level is that of dangerous or *hazardous* technology. The prime example of this is nuclear technology, even in this post-cold war era. The argument here is not merely against the production of nuclear weapons, but against all use of fission-based nuclear technology, because of the widespread consequences of any accident (Dumas 1999). The Fukushima nuclear power plant disaster following the great eastern tsunami of 2011 in Japan illustrates these concerns. Joy (2000) places robotics, genetic engineering and nanotechnology in this same category as well. More recently, an Oxford philosophy professor has warned that artificial intelligence could enslave humans (Bostrom 2014). Stephen Hawking, the well-known Cambridge theoretical physicist, also expressed such concerns shortly before his death in 2018.

The second level is one that is not necessarily dangerous for the entire human race, but where technology promotes *injustice*. For example, an infrastructure project is generally evaluated on the basis of a cost-benefit analysis. More sophisticated versions of such analyses would include social and environmental costs and benefits too. However, there could be a tendency for costs to be borne by one segment of a population and the benefits enjoyed by another. One of the most trenchant critiques of this disparity has been made by Booker Prize winning author Arundhati Roy (1999) in her article 'The Greater Common Good', a concept used by successive Indian politicians to persuade villagers facing the threat of displacement through the construction of large dams that their sacrifices were noble. In fact, she shows how the benefits accrue mostly to rich industrialists; while benefits such as power and irrigation that could serve a wider population are not fully realized because of siltation and salinization respectively. She raises serious questions about the wisdom of constructing large dams.

There are also the issues of ecological injustice, in that most decisions are taken on the basis of benefit for humans alone; and also of intergenerational injustice, since the welfare of future generations is often sacrificed for the sake of the present (Ferre 1995). In such cases technology may appear to be neutral; however, it could create the potential to exacerbate the injustice. At the same time however, technology has reduced injustice by bringing societies out of feudalism and improving standards of living. Similarly, technologies such as the internet have resulted in greater distribution of knowledge and hence power, though only to those having access to it (Feenberg 1999).

We come now to the third level, and with it to the less obvious influences of technology; this is the *sociological* influence of technology. Let us consider a few examples. Construction and transportation technology have created greater mobility and freedom; but have also contributed to the fragmentation of society, whether in the form of differences between suburbs and inner cities, or in the diminishing of extended family interactions. The construction of centralized facilities—such as a large power plants or transmitting stations—results in their security becoming very important. Societies with such centralized (as opposed to distributed) facilities will tend to be more securitized or even militarized, thus inadvertently reducing human freedoms in general. Engineers involved in such large projects would undoubtedly view themselves as contributing to social needs and sharpening their skills in the process (apart from the employment or remuneration they would enjoy); they may be completely blind to the securitization of society that could ensue. This is why engineers need, in the spirit of Heidegger, to be suspicious of the so called social benefits they are supposed to be delivering through their work.

Computer and communications technology have brought the world closer to us, but also made us a society of individuals, absorbed in one visual screen or other. Heidegger pointed out that changes in devices altered important life practices (Zimmerman 1990). For example, the change from writing to typing and then to word processing has changed the character of communication, making it less personal. Changes towards automation in the technology of domestic heating have reduced the level, not only of human involvement, but also of mutual participation. For example, when homes had to be heated by a log fire, many homes had just a single fireplace in the living room, in which all members of the family assembled, thus promoting family conversation. The advent of central heating however removed this need for the family to meet together. More recently, an MIT sociologist has described how the art of conversation has been undermined by the digital revolution (Turkle 2015). Technology's sociological impacts are not always obvious. When they are perceived however, what is more difficult is to decide whether to boycott such technologies or merely to mitigate their impacts.

The deepest level is the *psychological* influence of technology—not only through its artefacts and systems, but also in the emergence of a technological attitude. This has created a society where 'technique' is all important, as opposed to understanding (of phenomena) or even genuineness (in relationships), reflected in the growing number of 'how to' books—one of the earliest well-known examples is Dale Carnegie's (1936) *How to Win Friends and Influence People*. Another aspect of psychological impact is that we tend to value ourselves against our technological innovations. Shallis (1984) argues that the invention of the clock resulted in persons being judged for their efficiency, while that of the computer resulted in them being judged for their logical thinking. This threatens the value of spending time with others for developing relationships; and that of exercising of other forms of thinking, such as the imaginative or intuitive. While there may be some (or even great) benefits in technique, efficiency and logical thinking, there may be other aspects of our humanness that we lose in the process too. All this echoes Heidegger's concern that man himself has been shaped (i.e. 'enframed') by technology.

Consider finally Heidegger's concern that a thing should gather together 'earth and sky, divinities and mortals'; also that a work of art should 'set forth the earth' and 'set up a world'. Do we see this in today's technological artefacts? How can these categories be even interpreted for such artefacts? In the spirit of Feenberg (1999), can we make appropriate interventions rather than jettisoning technology altogether? Let us consider a computer and what it stands for as an example. The computer certainly sets up a world and generates a particular kind of community of mortals through electronic mail and the internet; it is weak however in setting forth the earth, because the computer and its attendant world are highly processed forms of raw material and used in very artificial work environments. Perhaps one way in which we can affirm both 'earth' and 'sky' is to use computers to reduce adverse impacts on the environment; this could take the form of reducing the use of paper, and miniaturization. Also, the move towards 'the internet of things', which envisages microprocessors embedded in both the natural and artificial environment, may bring a greater closeness to 'earth'; on the other hand, it may alienate us further from the earth, allowing the computer to mediate the earth to us rather than sensing it directly, even while on holiday or vacation.

The computer and its world appear to be deficient in 'divinities' too. This is because we try in our artificial environments to reduce uncertainty to a minimum—compare this with Heidegger's reflection on an agrarian world (Sect. 6.3 above), characterized by "anxiety as to the certainty of bread, the wordless joy of having once more withstood want, the trembling before the impending childbed and shivering at the surrounding menace of death". However, uncertainty and powerlessness of a different form may now be confronting modern society—not arising from nature but from the enormous complexity of the socio-technical systems that pervade us.

On the other hand, the very rationalism of the computer age could have been responsible for the advent of postmodernism, that can at one level be described as a search for transcendent meaning in the face of the bleak material commonplace. Coyne (1995) argues that our very entanglement with technology leads to a search for meaning beyond it. Also, while facing up to the meaninglessness or 'groundlessness' of modern technology, we could try to do 'little things' to alleviate it, as advocated by Heidegger (Zimmerman 1990)—even simply by adding personalized notes to circular letters sent through e-mail lists.

Questioning technology in practice is not an easy task. Not only does it involve questions such as socially acceptable levels of risk, but also issues of justice and values. There is also the need for a shared discourse and consensus, which is increasingly difficult to find today. While the treatment of these subjects is very brief here, the intention has been to argue that engineers should be part and parcel of this questioning process. One of the most important things to understand is that technology is not 'neutral'; the adoption of any given technology has implications for the levels of safety, justice, community and humanness that can either be promoted or diminished by it. As Heidegger (1977, p. 35) remarked, "Questioning is the piety of thought".

6.5 Ethics and Metaphor

We have already dealt with engineering ethics in Sect. 5.6. Heidegger (1997) gives us another dimension of ethics in describing our interaction with the world as one of 'care', albeit in the commonplace sense of taking care of things. All professionals owe a 'duty of care' to their clients. We could say therefore that Heidegger made a contribution towards ethics through his advocacy of questioning and his notion of care; and engineering ethics has indeed been treated as an ethic of care (Pantazidou and Nair 1999). The idea of care is associated more with the health profession, or with professions that have direct interaction with individuals. However, Tronto (1993) identifies attentiveness, responsibility, competence and responsiveness as the elements of care, and all of these are relevant for engineering. Attentiveness has to do with translating the client's requirements to engineering specifications. Responsibility also figures prominently in engineering, which is often project-oriented. Responsiveness has to do with ensuring that the client is 'delighted' (Blockley and Godfrey 2000), and also with learning through feedback. Perhaps the most relevant element for engineers is competence, through having specialized expertise combined with an awareness of the 'big picture' (Dias and Blockley 1995); and a willingness to consult other actors for aspects beyond their own expertise.

Also, just as Heidegger regarded poetry as an antidote to modern technology, Coyne (1995) argues that design should be guided more by metaphor than method. Snodgrass and Coyne (1992) suggest that models (or metaphors) used for design should be judged not only on the basis of their logical coherence, but also on the extent to which they are open ended and able to generate other metaphors. So, for example, the design of a house can be guided by the *metaphor* that "a house is a machine for living in". This creates open ended design that is creative, because there are many ways in which a house can be imagined as a machine. Contrast this to a *method* that says "a house should be designed so that spaces for complementary activities are placed adjacent to each other"—this will lead to a kitchen being placed next to a dining room and bathrooms next to bedrooms. Method is logical and rational, but lacks the creative potential of metaphor. The machine metaphor can also lead to the metaphor of 'mechanical' in the sense of 'without thinking', suggesting that using the house should be intuitive for its inhabitants. Metaphor based design can both reveal and conceal, a notion central to Heidegger—seeing the house as a machine reveals its operational features but probably conceals the aspect of safety; which could have been expressed in an alternative metaphor such as "a house is a safe haven". Engineering design may not be as open ended or amenable to metaphor as the architectural design described above. However, to see a truss as a plate for example (with non-essential areas removed), can generate rather novel truss configurations (Beghini et al. 2014).

6.6 Two Caveats

It is pertinent to sound two warnings at this stage. The first is to ensure that questioning technology does not lead to rejecting all of it, as sometimes espoused by anti-technologists. The call in this chapter is for engineers (i.e. technologists) themselves to be self-critical about the broader implications of technology. Such criticism should be constructive however, in the light of an overall mood of optimism regarding technology, as captured in the words of Popper (1999, p. 104) quoted in Sect. 2.2—i.e. about how technology contributed to the emancipation of women.

The second warning has to do with the use of Heidegger to advocate ethics. His association with Nazism could disqualify him as an ethicist. His view that the mechanization of agriculture was of the same essence as the holocaust (Collins 2000) severely erodes notions of justice. Furthermore, his existentialist philosophy is at the core nihilistic, for he said that there was no ultimate ground for our being (Dreyfus 1991); and it is not clear that ethics can be founded on such nihilism. On the other hand, there are many other frameworks that offer a more consistent and comprehensive rationale for technological ethics. The Judeo-Christian framework for example has been used to provide critiques of and guidelines for technology, with respect to both social justice and the natural environment (Conway 1999). However, despite some of these shortcomings, parts of Heidegger's writings do provide some very compelling bases for engineering ethics, largely through his warnings to be suspicious of technology.

6.7 Summary

- Heidegger affirmed traditional technology, *techne*, but was opposed to science based modern technology, in which everything (including man) is 'levelled down' to be a mere 'resource', i.e. an input to a process. In traditional technology on the other hand, an artefact was seen as a gathering together of a rich variety of causes, such as material, form, purpose and agent.
- Heidegger cautioned against treating technology as merely *neutral*, because things in the world both revealed and concealed their true nature. He advocated that we interpret the world with a spirit of *suspicion*.
- This spirit of questioning is something that engineers (who are agents of technology) would do well to emulate. Examples are given as to how we can question the increasingly deeper impacts of technology, ranging from its possibly hazardous nature, through its potential to create injustice, to its subtler sociological and psychological influences. The adoption of any given technology has implications for the levels of safety, justice, community and humanness that can either be promoted or diminished by it.
- Just as Heidegger considered poetry to be an antidote to the spirit of modern technology, the use of metaphor could also liberate design from its sometimes narrow rationalism and reductionism. In addition, Heidegger's notion of 'care' could be used as a basis for engineering ethics.

Acknowledgements Adapted by permission from Springer Nature Customer Service Centre GmbH: Springer Nature: *Science and Engineering Ethics, 9*(3), 389–396: Heidegger's relevance for engineering: questioning technology, by W. P. S. Dias, 2003.

References

L.L. Beghini, A. Beghini, N. Katz, W.F. Baker, G.H. Paulino, Connecting architecture and engineering through structural topology optimization. Eng. Struct. **59**, 716–726 (2014)

D.I. Blockley, P. Godfrey, *Doing It Differently: Systems for Rethinking Construction* (Thomas Telford, London, 2000)

N. Bostrom, *Superintelligence: Paths, Dangers, Strategies* (Oxford University Press, Oxford, 2014)

D. Carnegie, *How to Win Friends and Influence People* (Simon & Schuster, New York, 1936)

J. Collins, *Heidegger and the Nazis* (Icon Books, Cambridge, 2000)

R. Conway, *Choices at the Heart of Technology: A Christian Perspective* (Trinity Press International, Harrisburg, 1999)

R. Coyne, *Designing Information Technology in the Postmodern Age: From Method to Metaphor* (MIT Press, Cambridge, 1995)

W.P.S. Dias, Heidegger's relevance for engineering: questioning technology. Sci. Eng. Ethics **9**(3), 389–396 (2003)

W.P.S. Dias, D.I. Blockley, Reflective practice in engineering design. ICE Proc. Civ. Eng. **108**(4), 160–168 (1995)

H.L. Dreyfus, *Being-in-the-World: A Commentary on Heidegger's "Being and Time, Division I"* (MIT Press, Cambridge, 1991)

L.J. Dumas, *Lethal Arrogance: Human Fallibility and Dangerous Technologies* (St. Martin's Press, New York, 1999)
A. Feenberg, *Questioning Technology* (Routledge, London, 1999)
F. Ferre, *Philosophy of Technology* (University of Georgia Press, Athens, 1995)
M. Heidegger, *Poetry, Language, Thought* (trans: A. Hofstader). (Harper and Row, New York, 1971)
M. Heidegger, *The Question Concerning Technology and Other Essays* (trans: W. Lovitt). (Harper and Row, New York, 1977)
M. Heidegger, *Being and Time* (trans: J. Stambaugh). (SUNY Press, Albany, 1997)
B. Joy, Why the future doesn't need us. *Wired*, Ideas 04.01.00 (2000), https://www.wired.com/2000/04/joy-2/. Accessed 12 Dec 2018
C. Mitcham, *Thinking Through Technology: The Path Between Engineering and Philosophy* (Chicago University Press, Chicago, 1994)
M. Pantazidou, I. Nair, Ethic of care: guiding principles for engineering teaching and practice. J. Eng. Educ. **88**(2), 205–212 (1999)
K.R. Popper, *All Life Is Problem Solving* (Routledge, London, 1999)
A. Roy, The greater common good. Frontline **16**(11), 31 (1999)
M. Shallis, *The Silicon Idol: The Micro Revolution and Its Social Implications* (Schocken Books, New York, 1984)
A. Snodgrass, R. Coyne, Models, metaphors and the hermeneutics of designing. Des. Issues **9**(1), 56–74 (1992)
J.C. Tronto, *Moral Boundaries: A Political Argument for an Ethic of Care* (Routledge, New York, 1993)
S. Turkle, *Reclaiming Conversation: The Power of Talk in a Digital Age* (Penguin Press, New York, 2015)
M.E. Zimmerman, *Heidegger's Confrontation with Modernity: Technology, Politics and Art* (Indiana University Press, Bloomington, 1990)

Chapter 7
Is Knowledge Acquired by Thinking or Doing?

7.1 Being and Time

Martin Heidegger was a 20th century German philosopher who had enormous influence on shaping modern philosophical thought, and probably practical lifestyles too. He was part of the 'existentialist' school of philosophy, which stressed the importance of individual freedom and 'authenticity' (Dreyfus 1988). In other words, rather than making choices and performing actions according to some overarching theory (e.g. engineering method) or cultural practice (e.g. the current state of the art), we should do so in response to our unique experience and context—it is this that would give us freedom and make us authentic. It should be noted that experience and context are very important for engineering, whereas overarching theory (or generalization) and idealized models are the main focus of science. Heidegger's main preoccupation was with the question of *being*, which is the aspect of philosophy called *ontology* (see Sect. 2.1).

We show in this chapter how many aspects of Heidegger's ontology are particularly appropriate for and embodied in engineering. We also argue that Heidegger's philosophy can help engineers to understand their role and being; this includes his thoughts on the way that time impacts our way of being as humans. Heidegger is a very difficult philosopher to read, and is accused by some of being pretentious in his deliberate creation of new words. His alleged complicity with the Nazis during the second world war doesn't advance his cause either. Nevertheless, he appears to be highly relevant for engineers (Dias 2006). His major book is appropriately called *Being and Time* (Heidegger 1997).

P. Dias, *Philosophy for Engineering*, SpringerBriefs in Applied Sciences and Technology, https://doi.org/10.1007/978-981-15-1271-1_7

7.2 The Primacy of Practice Over Theory

One of the main thrusts of Heidegger's philosophy is the primacy of practice, or rather practices that we are socialized into, prior to any theoretical understanding. Heidegger approached the question of being from what he called 'the human way of being'. He did this because humans were the only beings who were concerned about their own being. He used the term *Da-sein* to denote this being. In addition to meaning 'the way humans are', this hyphenated German word can also mean 'being-there' and 'everyday human existence'. Heidegger argued that *Da-sein* was not a conscious subject, and that its way of being was 'being-in-the-world'; in other words, human beings always had the notion of a 'world', which meant a 'pre-theoretical' shared agreement in practices.

Also, subject-object distinctions were blurred in our everyday lives in the world. Dreyfus (1988), one of Heidegger's best exponents, gives the example of a person turning a doorknob to enter a room. In this very everyday act, he says, there is no conscious intention on the part of the person directed towards the doorknob, and hence no subject or object as such; rather, there is a seamless web of activity for the fulfillment of a purpose, in which both the person and doorknob are participants. Heidegger himself described this with the example of a carpenter using a hammer (Heidegger 1997, p. 65)—the terms in square brackets here and below are from a previous, more literal translation (Heidegger 1962) and the italics from the original.

> The less we just stare at the thing called hammer, the more actively we use it, the more original [primordial] our relation to it becomes and the more undisguisedly it is encountered as what it is, as a useful thing [equipment]. The act of hammering itself discovers the specific 'handiness' of the hammer. We shall call the useful thing's kind of being in which it reveals itself by itself *handiness* [*ready-to-hand*].

Once again this seamless web of everyday activity can be seen as not having any subject-object distinction; and the kind of knowledge required as being essentially practical—i.e. know *how* rather than know *what*. This is similar to Michael Polanyi's (1958) idea that a scientists had to 'interiorize' their tools of scientific exploration (see also Sect. 5.5). The awareness of a blind man exploring a cave with a stick is not of the stick but of the cave—since the stick becomes an extension of his arm; so also scientists 'interiorized' their theories in order to explore the world thereby (Polanyi 1966). Polanyi is however talking about intellectual tools (i.e. theories), but Heidegger about physical ones. Meanwhile, John Dewey argued that the act of knowing was technological in nature, employing internal tools such as ideas and language; in fact, it is technological words such as 'construction' (e.g. of a theory or argument) that are often used to describe thought processes (Hickman 1990).

Getting back to Heidegger's carpenter, the hammer could become *conspicuous* (if it was too heavy), *obtrusive* (if it could not be found) or even *obstinate* (if the head came off the hammer). This then forced the carpenter to pay attention to the hammer. There was now revealed a subject with a mental content (i.e. the carpenter) and a thing that was seen as an object (i.e. the hammer). The context however was still important, and the carpenter was still concerned about the hammer not as an isolated object, but one to be used for a purpose. Heidegger (1997, p. 68) referred to this mode of being of the hammer as *unhandiness* [*un-ready-to-hand*]. He also said that this un-readiness-to-hand pointed to two other ways of being, namely *present-at-hand* and *just-present-at-hand* (Heidegger 1997, p. 69):

> When we discover its unusability, the thing becomes conspicuous. *Conspicousness* presents the thing at hand in a certain unhandiness [un-ready-to-hand].
>
> With this obstinacy the objective presence [present-at-hand] of what is at hand makes itself known in a new way as the being of what is still present and calls for completion.
>
> As a deficient mode of taking care of things, the helpless way in which we stand before it discovers the mere objective presence [just-present-at-hand] of what is at hand.

These passages reflect greater degrees of subject-object distinction and greater focus on the 'objective' properties of things. They indicate however that these ways of being are derived from the 'ready-to-hand' way of being, and are actually contained within it. These derivative ways of being show up when the primordial way of being experiences 'breakdowns'. Heidegger insisted that the analytical isolation of *fundamental* properties of objects by detached subjects was a way of being that was derived from a more *primordial* way of being, where a seamless subject-object continuum achieved purpose through practical action. Turk (2001a) argues that objects, or at any rate their properties, are human constructs and not things that have existence independent of ourselves; he also says that the original Latin *ob-iectum*, meaning 'thrown in front of us', clearly suggests observer participation in objects.

In other words, to say that a carpenter's hammer is "something that is useful and appropriate for carpenters to drive nails into timber" is a richer and more primordial description than saying that a carpenter's hammer is "an object of around 20 oz (0.57 kg) weight made of a metal head firmly jointed to a wooden handle". Note that the former description is also capable of being physically manifested in many ways—it is a 'higher' level functional description, rather than the 'lower' level physical one. Heidegger is saying that the very need for the latter analytical description has arisen because humans in the past have experienced 'breakdowns' in the seamless practical action of the first description—breakdowns caused either by picking up a heavier (or lighter) than usual hammer, or having the head come off the handle. Another way of saying this is that we do not get our mental picture of the world by aggregating the so called fundamental properties (e.g. material, weight, connectivity) of all its components (e.g. hammers, carpenters etc.) and their inter-relationships. Rather we get this mental picture by living in the world and being socialized into its practices through our various practical interactions. The need for fundamental properties arises only when we need to understand why our everyday experience has been disrupted, and how we can remedy the disruption to prevent it in future.

Heidegger's philosophy could serve as an intellectual platform for combating the feelings of inferiority and lack of status that many engineers worldwide experience in a culture (still heavily influenced by Plato and ancient Greece) that values analysis more than synthesis, and theoretical knowledge more than practical intelligence (see Sect. 2.2). Patrick Nuttgens (1988) has argued that children first learn about the world by practice before they acquire a theoretical framework; and that technical education should reflect this (see also Sect. 2.4). One easy implementation of this in engineering undergraduate programs could be to have students performing a set of experiments related to a given subject before learning the theory related to it.

7.3 The Engineer's Existential Role

Another important idea from Heidegger is that of our existential (or experiential) situation. This is defined by what could be called a 'web of references or relations'. Heidegger said that we encountered things in the world not as isolated objects, but as *equipment* having a purpose. The Greek word for things is *pragmata*, from which we get the English word 'pragmatic', and conveys this very idea (Heidegger 1997). However, we never encountered equipment in isolation; rather, we encountered them in relationships to other equipment, such as pen and paper (and table and lamp and so on). It was within such an equipmental whole that the equipment had meaning. The equipment was employed towards a purpose, but this itself was subservient to *Dasein's* self-interpretation (Heidegger 1997). Dreyfus (1991, p. 92) gives an example of Heidegger's relationships by an example where he says:

> I write on the blackboard *in* a classroom ('where-in' or practical context), *with* a piece of chalk ('with-which' or item of equipment), *in order to* draw a chart ('in-order-to' conveying purpose) *towards* explaining Heidegger ('towards-which' or goal), *for the sake of* my being a good teacher ('for-the-sake-of-which' or final point). (Italics from original; parentheses by present author based on original)

Heidegger (1997) called this a structure of *reference* or *serviceability*; note that the latter is an engineering term too, that describes regular (everyday) functioning. Each person comes across this existential situation in what Heidegger called the 'average everyday' world. Although Heidegger's philosophy of 'everydayness' can be made relevant for persons having any profession or even none, his ideas are especially apt for describing engineers. Table 7.1 presents engineering correlates for some Heideggerian terms and summarizes the discussion below.

The first six rows in Table 7.1 relate to the structure of reference described by Heidegger. Heidegger's practical context can be compared to the context for an engineering project. Such engineering projects, whether design, fabrication or maintenance, will almost invariably use equipment of one sort or other (whether computers or cranes). Purpose (towards which the 'equipmental whole' is deployed) is also very pertinent to engineering, which is a very pragmatic discipline oriented towards specific objectives. Needless to say, any project will have a final goal, but an engineer will also be concerned about his career, which transcends the project in some ways. More than most other professionals, an engineer is aware of his socialization. In probably no other profession do so many professionals work together in a single organization. Doctors, lawyers and architects generally practice their professions very individualistically. In addition, their contact with the public tends to be on a one-to-one basis. Engineers on the other hand make contributions that affect the entire public or large

Table 7.1 Engineering correlates of some Heideggerian terms (after Dias 2006)

Heideggerian term	Explanation	Engineering correlate
'Where-in'	Practical context	Context for project
'With'	Item of equipment	Piece of equipment
'In order to'	Purpose	Objective
'Towards'	Goal	Goal of project
'For the sake of'	Final point	Career
'Being-with'	Socialization	Public, employer, fellow engineers
'Thrownness'	Given context	Situation at hand
'Understanding'	Future possibilities	Alternatives
'Discourse'	Articulation	Interpretation of situation
'Falling prey'	Absorption in everydayness	Standard approach to work
'Choose'	Choice	Selection of strategy
'Resoluteness'	Tenacity	Sticking to a plan
'Temporality' (future)	Mortality	Creating 'monuments'
'Temporality' (past)	Cultural history	Case histories
'Situation' (present)	Moment of choice	Immediate context
'Temporality' (future)	Mortality	Creating 'monuments'
'Temporality' (past)	Cultural history	Case histories
'Situation' (present)	Moment of choice	Immediate context

sections of it, but remain invisible to this public and have little or no contact with them. However, practising engineers are very aware of 'being-with' other engineers and professionals during their careers (Davis 1998).

The last nine rows in Table 7.1 relate to the way that a person would live existentially in this 'everyday world', the first six covered in this section and the last three in the next. Heidegger (1997) used the word 'care'—in the sense of taking care of things—to describe *Da-sein's* stance towards the world, or involvement in it. Such care was dependent on three factors. The first was 'attunement' (of which mood was an important aspect), and related to the *past*. This was also called 'thrownness', in the sense that *Da-sein* was always thrown into a particular context, both social and personal. The second was 'understanding', and related to the *future*. Understanding was considered a 'standing-out' into future possibilities; this was always constrained by thrownness, but a certain range of possibilities was always available to *Da-sein*. The third factor of care was 'discourse', which related to the *present*. This has a significance wider than a purely linguistic one. Another term that Heidegger used was 'articulation', which has both ontological and linguistic significance. Something that is articulated has its connections and joints disclosed—e.g. an 'articulated' truck. It was this articulation that enabled us to interpret and give meaning to our world.

In its everyday involvement in the world, *Da-sein* was described by Heidegger as 'falling prey' to it. Although the idea of falling has theological overtones, Heidegger meant by this value free term that *Da-sein* was absorbed in the 'average everydayness' of the 'one' (as in "One should respect one's teachers" or "One does not use a hammer to break an egg"). So although average everydayness was considered to be the primordial way of being, it was essentially a 'concealment' (see Chap. 6 too), because it cut *Da-sein* off from its potentialities. *Da-sein's* response to falling prey was called '*angst*', which could be translated 'anxiety', although Heidegger's translators leave the word untranslated. The theological counterpart would be 'guilt', but Heidegger had a slightly different concept in mind. It was *angst* that suggested to *Da-sein* the potentialities for its 'ownmost' way of being, in contrast to the being of the 'one' or that of average everydayness. In short, all of us are socialized into a world that does things in particular ways. We cannot avoid being like everyone else in most cases, but we all suspect that there are some things that can be done "in my own way", thus giving us freedom and authenticity; in other words, we have *angst* until we choose with tenacity to act in ways that are "true to ourselves".

Engineering can be seen as a problem solving discipline that is largely project based. A project will have some unique features that require context sensitive action. As such, an engineer can be seen as being thrown into a 'situation at hand'. Within such a context, any project will have a limited but real range of (future) outcomes that can be achieved. The engineer has to interpret his situation in the light of the state of the art of his profession and the range of possibilities or alternatives available to him. He could perhaps be merely carried along the stream of the standard approach to work. The authentic engineer however will on occasion select an appropriate strategy for himself, always of course in the "sober understanding of the basic factical possibilities" (Heidegger 1997, p. 286) and in relation to other engineers and 'actors'; but then stick to his plan, even though it may be a little away from the beaten track. This choice and resoluteness is reminiscent of a military ethos, which has been described as one of the well-springs of engineering (Davis 1998).

The above relevance of Heidegger's notion of 'thrownness' to engineering practice has been cogently argued by Turk (1998, 2001a). He also discusses the 'blindness' that would be experienced if engineers rely on a *generalized* conceptual model of either the product or process they are involved in, rather than recognizing their 'thrownness' in a given context, with its *particular* idiosyncrasies (Turk 1998, 2001a). Engineers do use models, whether physical, mathematical or conceptual, especially when faced with 'breakdowns'. They must ensure however that the models represent context as faithfully as possible, and recognize that model outputs have also to be judged in the light of specific contexts. We have come across similar ideas about engineering models in Sect. 4.5 too. In other words, engineers are committed primarily to 'being-in-the-world' rather than to 'being-in-a-model-of-the-world' (Turk 2001b).

Very often engineering boils down to decision making under uncertainty (including incomplete knowledge), in most cases under time constraints too. Engineers have to make do with the little information they have in the limited time available. This is why engineering *judgement* is called for, and it is Heidegger (as opposed to say Plato) who gives respectability to such a mode of decision making and form of rationality. Aristotle (2000) would have called this *phronesis* (practical wisdom for action) and recognized its importance; it is a great pity for engineers that he considered *sophia* (theoretical wisdom) to be superior. There are current philosophical trends however that espouse a higher intellectual stature for *phronesis* (Long 2002).

7.4 Engineering in Time

We have seen how the dimensions of care are time-related. In addition, Heidegger (1997, pp. 243–5) identified a deeper set of time-based factors that affected *Da-sein*, and promoted individuation, or *authenticity*. The future-based factor was *death*, which Heidegger described as being 'ownmost', non-relational (i.e. to be faced individually), not to be bypassed, certain and indefinite (with respect to when it will take place). An appreciation that *Da-sein* was in fact 'being-towards-death' would encourage the pursuit of one's particular potentialities. The present-based aspect of temporality was what Heidegger (1997) called the '*situation*'. It was the current moment in which *Da-sein* had to choose an authentic action or direction in contrast to what 'one' would have done on average. It was for this authentic choosing that resoluteness was required.

The past-based factor was the *cultural history* into which *Da-sein* had been socialized, starting from birth. The influence of temporality was thus extended from *Da-sein's* birth to death. At first sight, this cultural history would suggest an averaging tendency as opposed to an individuating one, since social norms are based on such cultural history. Heidegger referred however to the historicity of being, in that differing ways of being human showed up in different times and cultures. For example, heroic behavior was revealed in ancient Greece, and saintliness in Christian culture (Dreyfus 1991). So *being* had to be interpreted in the light of a time horizon. However, present day *Da-sein* could obtain role-models for emulation from historical periods other than its own. In the same sense, Heidegger (1977) also said that history was a critique of the present.

The aspect of temporality has particular relevance to engineers. Apart from the state of the art, they will have the history of the profession to guide their choices. This highlights the importance of case histories for engineers, both in their initial and continuing education (Dias and Blockley 1995; Dias 2014). Examples from the past can bring fresh insight into a state of the art that has fallen into a 'rut'. Another source of the quest for originality that drives many 'builders', whether rulers, owners or building sector professionals, may be an appreciation of their mortality, and a desire to leave behind unique 'monuments'.

Time, on a day to day basis, is also very relevant to an engineering project. Although a project schedule is based on calendar days, it is the inter-relationships between activities, their pre-requisites and durations that are important. Each project would therefore have a string of activities that constitute its 'critical path'; it is this time-line that is of consequence, rather than the sequence of passing days. As we see in the last row of Table 7.1, every moment in a project is a 'now-that', an immediate context that has to be evaluated for possible corrective or opportunistic action. Working 'overtime' is therefore very common in engineering projects. It is not the pre-specified length of a working day that matters, but whether the work scheduled for the day has been completed.

Because of the importance of time in engineering projects, it is a resource that is occasionally traded off with others, such as accuracy, and even quality. A design engineer who is pressed for time may not be able to use refined analytical techniques; he will however 'play safe' and his design will probably involve a higher margin of safety (with some increase in cost too). A construction engineer who is fighting a deadline may be tempted to take 'short-cuts' on some aspects of construction quality, especially if the deviation has only a small impact on the final product (Dias 1997). At any rate, the time dimension, together with its perceived importance in engineering projects, increases the amount of engineering judgement required.

7.5 From Average Everydayness to Existential Pleasure in Engineering

Given that the engineering state of the art can often result in a standard approach to work, engineers may at times experience a certain lack of fulfillment or even boredom in their work. What is worse, it may be perceived that their work can be performed by those without engineering qualifications or credentials. In some societies, less qualified persons, such as drivers of train engines, do indeed bear the title 'engineer'. Employers who wish to cut costs could substitute engineers with technicians, but still bestow the job title of 'engineer' on them. This would be inconceivable in the medical and legal professions. In some cases, a technician (with less formal education) may even perform better at routine tasks. This can lead to a feeling of inferiority for engineers. It can also lead to questioning the value of a formal, rigorous engineering education. The gap between engineering education (largely theoretical and mathematical) and practice (largely practical and empirical) is widely acknowledged (Dias and Blockley 1995), and in probably no other profession is that gap so manifest; but the way to bridge it is by no means clear.

Heidegger's moves away from 'average everydayness' can provide a way towards the resolution of the above dilemma. There are two ways in which such a move can take place. The first is precipitated by 'breakdown'. The carpenter in Heidegger's example is involved in very practice-based unconscious hammering only when nothing is wrong. However, when there is a breakdown in this everydayness, say when the hammer is too heavy, the carpenter will have to resort to 'mentality' and study properties such as the weight of the hammer object; or if the head comes off the handle, once again he will have to give careful attention to the properties of the joint, in order to solve the problem. In fact, Heidegger considered that scientific observation and reflection took place precisely at such breakdowns.

This then would be the engineers' justification for their theoretical training. Although they may be using a standard approach to engineering practice in the normal course of events, they will need a bedrock of theoretical knowledge to fall back on when faced with problems that intrude into their work. Many professional engineering organizations, in the process of admitting engineers to full membership after a period of work-based training, are interested in finding out about problems encountered during the engineer's work, and how 'engineering first principles' were used to overcome them. It should be noted that this scientific analysis need not be confined to physical entities; they can also be directed at the social context in which engineering takes place, e.g. via the sociological analysis of failures (Turner and Pidgeon 1997).

Furthermore, especially in civil engineering practice, every project has some degree of unpredictability that causes deviation from its design or planning, particularly when the project is carried out in a complex natural and social environment. These problems can range from a bored pile excavation that may encounter a hard stratum before reaching the desired rock layer (causing excessive durations of boring), to a ready mixed concrete truck that may get delayed by traffic (casting doubts about the workability of the concrete). In spite of measures such as quality assurance—where it is sought to guarantee quality by carefully controlled work practices—engineers on site will often be forced to make judgements on the best course of action under non-ideal conditions (or breakdowns). For this, they will need recourse to deeper knowledge (Dias 1997).

Heidegger's other move from everydayness was in his description of choosing one's 'ownmost potentiality' rather than 'falling prey' to the way that things would normally be done. So, an engineer could display his ability to do more than the 'run of the mill', by asking whether even a routine activity could be done better—within practical limits of course, the importance of which was underlined by Heidegger too. Heidegger then gives us the rationale for self-actualization even in the midst of the routine. Florman (1994), in his book on *The Existential Pleasures of Engineering*, points to the self-actualization that comes from the work of engineers that certainly transcends average everydayness.

Another concept used to describe this is Schon's (1983) 'reflective practice', which has been applied to engineering as well (Blockley 1992; Dias and Blockley 1995). The modern novelist Robert Harris' (2004) well researched book *Pompeii* explores the experiences of a hydraulics engineer in Roman times; an engineering reflection on the novel (Dias 2010) highlights some of the above aspects of an engineering outlook to practice. Broome and Peirce (1997) use the term 'heroic' to describe what engineers should aim for, and be educated for, in order to "venture forth from the world of common day", both to make decisions under incomplete information and also to make their profession a 'caring' one—see also Sect. 6.5.

Heidegger was also concerned however, about the way that the true nature of things was concealed by our 'average everydayness', and his approach to philosophy was one of deep questioning. Among the many aspects of 'being' questioned by Heidegger (1977) were those of science and modern technology, which he thought reduced everything (including man) to the level of a mere 'resource'. On the other hand, he did affirm traditional technology, which he considered similar to poetry, which in turn he commended as an antidote to modern technology (Heidegger 1971). These aspects have been dealt with in detail in Chap. 6, which advocates for this spirit of questioning among engineers (who are the agents of technology); and also suggests that the use of metaphor could liberate design from its often narrow rationalism (Snodgrass and Coyne 1992).

In closing we can say that although Heidegger's stated preoccupation was with *being*, it appears that much of his writing in fact focuses on *doing*. In this respect too therefore, he has much relevance for engineering. Both Heidegger (1997) and Polanyi (1958) can be seen as advocating an 'instrumentalist' epistemology—i.e. the notion that we arrive at knowledge about the world through action as well, rather than by 'pure thought' alone. We return to both these philosophers again at the beginning of Chap. 8.

7.6 Summary

- In a cultural milieu where scientific understanding is prized over empirical action, resulting in engineers having self-doubt about their worth and status, Heidegger could serve as a patron philosopher for engineers, emphasizing as he does the primacy of practice over theory and also that of given contexts over idealized models.
- Heidegger addresses the gap between engineering education (generally highly theoretical, requiring analysis) and engineering practice (often very empirical, requiring judgement), by defining the relationship between *know how* (used during 'average everydayness') and *know what* (used during 'breakdowns').
- Heidegger clarifies the importance of the temporal element for engineers—e.g. learning from past case histories in addition to being guided by the state of the art; seeing our mortality as a spur to creativity; and seeing the present moment as always presenting opportunities for corrective or opportunistic interventions.
- Heidegger suggests two possibilities for moving away from 'average everydayness' in order to achieve our own *authenticity* and freedom. They are the use of theoretical knowledge during 'breakdowns' and the use of reflective practice in the continual succession of 'current moments'.

Acknowledgements Adapted by permission from Springer Nature Customer Service Centre GmbH: Springer Nature: *Science and Engineering Ethics, 12*(3), 523–532: Heidegger's resonance with engineering: the primacy of practice, by W. P. S. Dias, 2006.

References

Aristotle, in *Nicomachean Ethics*, ed. by R. Crisp (Cambridge University Press, Cambridge, 2000)
D.I. Blockley, Engineering from reflective practice. Res. Eng. Des. **4**, 13–22 (1992)

T.H. Broome, J. Peirce, The heroic engineer. J. Eng. Educ. **86**(1), 51–55 (1997)
M. Davis, *Thinking Like an Engineer: Studies in the Ethics of a Profession* (Oxford University Press, New York, 1998)
W.P.S. Dias, Sensitivity and substitutability in concrete construction. Asia Pac. Build. Constr. Manag. J. **2**(2), 32–34 (1997)
W.P.S. Dias, Heidegger's resonance with engineering: the primacy of practice. Sci. Eng. Ethics **12**(3), 523–532 (2006)
W.P.S. Dias, *Pompeii* by Robert Harris: an engineering reading. ICE Proc. Eng. Hist. Herit. **163**(4), 255–260 (2010)
P. Dias, The disciplines of engineering and history: some common ground. Sci. Eng. Ethics **20**(2), 539–549 (2014)
W.P.S. Dias, D.I. Blockley, Reflective practice in engineering design. ICE Proc. Civ. Eng. **108**(4), 160–168 (1995)
H.L. Dreyfus, Husserl, Heidegger and modern existentialism, in *Great Philosophers: An Introduction to Western Philosophy*, ed. by B. Magee (Oxford University Press, Oxford, 1988), pp. 252–277
H.L. Dreyfus, *Being-in-the-World: A Commentary on Heidegger's "Being and Time, Division I"* (MIT Press, Cambridge, 1991)
S.C. Florman, *The Existential Pleasures of Engineering*, 2nd edn. (St. Martin's Press, New York, 1994)
R. Harris, *Pompeii* (Arrow Books, London, 2004)
M. Heidegger, *Being and Time* (trans: J. Macquarrie, E. Robinson) (SCM Press, London, 1962)
M. Heidegger, *Poetry, Language, Thought* (trans: A. Hofstader) (Harper & Row, New York, 1971)
M. Heidegger, *The Question Concerning Technology and Other Essays* (trans: W. Lovitt) (Harper & Row, New York, 1977)
M. Heidegger, *Being and Time* (trans: J. Stambaugh) (SUNY Press, Albany, 1997)
L.A. Hickman, *John Dewey's Pragmatic Technology* (Indiana University Press, Bloomington, 1990)
C.P. Long, The ontological reappropriation of *phronesis*. Cont. Philos. Rev. **35**, 35–60 (2002)
P. Nuttgens, *What Should We Teach and How Should We Teach It?: Aims and Purpose of Higher Education* (Gower Publishing Company, London, 1988)
M. Polanyi, *Personal Knowledge: Towards a Post-critical Philosophy* (University of Chicago Press, Chicago, 1958)
M. Polanyi, *The Tacit Dimension* (Doubleday & Co., Garden City, 1966)
D.A. Schon, *The Reflective Practitioner: How Professionals Think in Action* (Temple Smith, London, 1983)
A. Snodgrass, R. Coyne, Models, metaphors and the hermeneutics of designing. Des. Issues **9**(1), 56–74 (1992)
Z. Turk, On theoretical backgrounds of CAD, in *Artificial Intelligence in Structural Engineering*, ed. by I. Smith (Springer, Berlin, 1998), pp. 490–496
Z. Turk, Phenomenological foundations of conceptual product modeling in AEC. Int. J. Artif. Intell. Eng. **15**, 83–92 (2001a)
Z. Turk, Multimedia: providing students with real world experiences. Autom. Constr. **10**(2), 247–255 (2001b)
B.A. Turner, N. Pidgeon, *Man-Made Disasters*, 2nd edn. (Butterworth Heinemann, Oxford, 1997)

Chapter 8
Can Practice Based Knowledge Be Formalized?

8.1 Practice Based Knowledge

Theoretical knowledge has been prized in academic institutions at least since the scientific revolution. The philosophical underpinnings for this, in the form of privileging the intellectual over the practical, have come from Descartes; recall that he put theory over observation with his dictum "*Cogito, ergo sum*" ("I think, therefore I am"). The deeper roots for a bias towards theory over practice go back to Plato himself. Plato said that what we see as *real* objects were mere copies of an *ideal* object; ideal objects are not real, so simplifying assumptions can be made to develop 'neat' theories about them—e.g. Boyle's Law which states that gas pressure is inversely proportional to volume holds only if the volume of the gas molecules is ignored. In engineering, this focusing on theoretical knowledge has caused a gap between academic training and professional practice. Engineering students at university solve differential equations, often using finite element analysis. Engineering practice however is mostly about using judgement and experience to tackle new problems (Dias and Blockley 1995). At the same time, many fields of engineering have craft based origins (Dias 2002); and this has given rise to a rich collection of heuristics or 'rules of thumb' that can supplement (or even supplant) the complex analysis.

Practice based knowledge however has not acquired the same 'respectability' as theoretical knowledge in academic institutions. Theoretical knowledge has *generality*, and applicability in a wide range of contexts; while practice based knowledge is seen merely as a collection of *particular* pieces of information. At the same time, theoretically trained engineers who spend their engineering careers doing routine tasks based on heuristic rules could question the value of their training and role—e.g. as to how it is different to that of a craftsman (Sect. 2.1). One reason for the above is that there are no formalizing principles for practice based knowledge, such as provided by the scientific method and mathematics for theoretical knowledge. It has been proposed that systems thinking frameworks and artificial intelligence (AI) techniques can provide formalizations for practice based knowledge at the conceptual and technical levels respectively (Dias 2002). In addition to such formalizations

P. Dias, *Philosophy for Engineering*, SpringerBriefs in Applied Sciences and Technology, https://doi.org/10.1007/978-981-15-1271-1_8

however, some philosophical arguments are required to counter the stranglehold that Plato and Descartes wield over our intellectual milieu.

This chapter focuses on the philosophers Michael Polanyi and Martin Heidegger. It demonstrates that Polanyi's epistemology (theory of knowledge) and Heidegger's ontology (theory of being) have the potential for placing practice based knowledge on a sound intellectual footing. It then explores two different categorizations of practice based knowledge, namely the *historical* (structured) and *horizontal* (unstructured). Finally, it gives some examples of problems that are not so amenable to theoretical knowledge; and demonstrates how Artificial Intelligence (AI) techniques can be used to capture, structure and process the practice based knowledge related to those problems (see also Dias 2007). We can, somewhat loosely, associate capturing with the conversion of data to information; structuring with the transformation of information to knowledge; and processing with the application of knowledge (in a new situation) to generate wisdom.

8.2 Michael Polanyi: Tacit Knowing

One of Polanyi's main contributions to epistemology was the idea of tacit knowing; an important book of his is titled *The Tacit Dimension* (Polanyi 1966). A key aspect of tacit knowing was that it attended from *particulars* to a *whole*. Polanyi used the example of recognizing a face to illustrate this—we use our *subsidiary* (or 'background') awareness of the features in order to achieve *focal* (or 'targeted') awareness of the face. The particulars were not to be focused on, but 'seen through', like using a pair of spectacles. To focus on the spectacles would mean that we cannot use them to see anything else (Prosch 1986). Similarly, directing our attention at the isolated features of a face would destroy the act of recognition. This focal recognition of wholes had similarities to Gestalt-type awareness (Polanyi 1958), where the whole 'falls into place', when deliberate attention is not paid to the particulars. Such subsidiary awareness of particulars meant that they could not be fully specified (Polanyi 1966). This is why Polanyi (1966, p. 4) said that "we can know more than we can tell". This 'from-to' process of acquiring knowledge could not be made explicit either; in other words, the path from particulars to whole is not reversible (Polanyi 1958). It is this *unspecifiability* of particulars and *irreversibility* of knowing that constitutes the tacit dimension (Polanyi 1966, p. 18):

> Scrutinize closely the particulars of a comprehensive entity and their meaning is effaced, our conception of the entity is destroyed. Such cases are well known. Repeat a word several times, attending carefully to the motion of your tongue and lips, and to the sound you make, and soon the word will sound hollow and eventually lose its meaning. By concentrating attention on his fingers, a pianist can temporarily paralyze his movement. We can make ourselves lose sight of a pattern of physiognomy by examining its several parts under sufficient magnification.

This movement from particulars to whole can be called *emergence*—knowledge emerges as we move from parts to whole. Polanyi also talked about emergence within entities or objects—new properties that are governed by new laws would emerge as we moved up a *hierarchy* of entities. The term 'holon' (Koestler 1967) has been used to describe entities that both comprise 'lower' entities, and are parts of 'higher' ones. He used the example of making a speech, in which he identified five hierarchically arranged elements, each of which had organizing principles governed by various laws (Polanyi 1966). So a voice (obeying the laws of phonetics) would be required to generate words (consistent with a lexicography); which in turn would be used to create sentences (governed by grammar) that had to be delivered in an appropriate style (according to the rules of stylistics) to fit an overall composition (that would be judged by literary criticism). The laws governing lower level entities would need to be satisfied or obeyed in order to achieve a higher level entity successfully—e.g. a grammatically correct sentence would be useless if words with incorrect meanings were used to form it. However, the principles controlling the higher level entity could not be fully derived from the laws governing the lower one, because they had organizational principles of their own—e.g. an engaging style of delivery may not result in a successful speech, if it was not composed in a coherent way.

Using the example of a machine, Polanyi (1958) stated that we would not be able to arrive at the *purpose* of the machine, even if we obtained a complete physico-chemical description of it. Purpose was an operational principle at the higher level of engineering, and required an engineering description. An engineering description could have many physico-chemical manifestations. The physico-chemical description would however define the conditions under which the machine could function; and explain the cause of any machine failure. The lower level laws could thus explain the failure of higher level operational principles, but not their success.

Polanyi (1969) also argued that knowledge involved skill. It means that knowing is an *active* process, requiring intelligent *effort*, as opposed to the passive perception of phenomena. It also means that there is a difference between knowing 'what' and knowing 'how'; and that the former is embedded in the latter (Polanyi 1958). Thus, there is an indefinable component in our knowledge that cannot be transmitted by propositions and statements alone. This is particularly evident in technological knowledge, and more so in its craft based elements, where apprenticeship within a tradition is essential for the passing on of skills. Polanyi used the example of cabinet makers; but he clearly meant apprenticeship to be applicable to scientists too—we know that every research student at a university learns to do research not primarily by following research methodology courses, but by being 'apprenticed' to an experienced academic supervisor.

Schon (1983) called this 'knowing in action', but also called for 'reflection in action', which involved an intimate interaction with one's self and context, and with others; his 'reflection *on* action' was to be done after acting, and can be called 'learning'. So there is increasing mentality as we progress from 'knowing in action' through 'reflection in action' to 'reflection on action'. He contrasted all of this 'reflective practice' (i.e. an aspect of practice based knowledge) with 'technical rationality', which paid selective inattention to all aspects of problems that could not be theoretically formulated.

8.3 Martin Heidegger: Pre-theoretical Shared Practices

We have seen in Sect. 7.2 that one of the main thrusts of Heidegger's (1997) philosophy is the primacy of practice, or rather practices that we are socialized into, prior to any theoretical understanding; in other words, human beings always had the notion of a 'world', which meant a 'pre-theoretical' shared agreement in practices. Also, subject-object distinctions were blurred in our everyday lives in the world. So, whether in the case of a person turning a doorknob to enter a room, or a carpenter using a hammer to drive in a nail, both persons and objects are involved in a seamless web of activity for the fulfillment of a purpose (Dreyfus 1988; Heidegger 1997). Heidegger insisted that the (analytical) isolation of *fundamental* properties of objects by detached subjects was a 'way of being' that was derived from a more *primordial* way of being, where a seamless subject-object continuum achieved purpose through practical action. In other words, all so called 'objective' properties had some relation to purpose in everyday life; otherwise they would not have been 'shown up'. We have explored this in detail in Sect. 7.2 (also Dias 2006).

The flip side of this derivative way of being associated with science is that scientific representation can never capture the totality of the world (Heidegger 1977). Where cognitive modelling was concerned, Heidegger criticized the symbolic representation of entities because it sought to 'free' objective properties of things by stripping away their contextual significance, and then to reconstruct a meaningful whole by adding further meaningless elements (Heidegger 1997, p. 82):

> The referential context that constitutes worldliness as significance can be formally understood in the sense of a system of relations. But we must realize that such formalizations level down the phenomena to the extent that the true phenomenal content gets lost, especially in the case of such 'simple' relations as are contained in significance. These 'relations' and 'relata' of the in-order-to, for-the-sake-of, the with-what of relevance resist any kind of mathematical functionalization in accordance with their phenomenal content. Nor are they something thought, something first posited in 'thinking', but rather relations in which heedful circumspection as such already dwells. (Quotation marks from original)

In other words, the holistic, context dependent way in which we encounter the world could not be represented. We would be trying to exchange a '*presencing*' of the world with a mere '*re-presentation*'. Our own skills of cognition too could not be captured by a predicate calculus. These skills of cognition would include our embodiment in physical bodies, as argued by Mearleau-Ponty (Dreyfus 1988) and Johnson (1987, 2007). At any rate, Heidegger's ontology turns Descartes' epistemological motto of "I think, therefore I am" around completely. To Heidegger, *being* preceded *thinking*; in other words, "I am, therefore I think", and this *sum* ('I am') too should read 'I-am-in-the-world' (Heidegger 1997).

8.4 Categories of Practice Based Knowledge

Both Polanyi and Heidegger are good advocates for the importance, and indeed primacy of practice based knowledge. How then can this knowledge be formalized and categorized? We have said before that AI can possibly provide a formalization for practice based knowledge. Within AI, Minsky (1991) has distinguished between cognitivist and connectionist approaches. The cognitive approach is epitomized by expert systems (Hayes-Roth et al. 1983). Here, the knowledge is made explicit by eliciting it from an expert, generally in the form of production rules (i.e. if…then… relationships). Uncertainty can also be built into the system. Once the knowledge base is thus prepared, facts concerning a new problem situation will trigger certain rules, and result in a diagnosis or decision. The triggering or firing of rules is governed by what is called the 'inference engine' of the expert system. One of the important features of expert systems is that the rationale for arriving at the end result is also made available to the user.

On the other hand, the connectionist approach does not force experts to formulate their knowledge in the form of rules. All it requires is that they codify their experience in the form of case histories. The computer then discovers patterns that even the experts may not have been aware of in their decision making; this is called the training phase. It can also predict the action that will be taken by an expert, if given the parameters that define a new problem situation. In this approach, the knowledge is implicit and no explanations are given to the user (Coyne 1990). An Artificial Neural Network probably best epitomizes this approach. Polanyi and Heidegger refer to the difficulties of 'specifiability' and representation respectively. As such, they would seem to be less in favour of and even quite opposed to the cognitivist approach. On the other hand, the connectionist approach of pattern recognition—including its feature of being unable to give explanations to users—resonates very strongly with Polanyi's tacit knowing, reflecting his notion of 'irreversibility'.

Another categorization of practice based knowledge is that of the *historical* versus the *horizontal*. Discipline related information, comprising engineering science theories and codes of practice, can be termed *vertical* knowledge, and we shall not refer to it further. In addition, during a given design project, there will be *horizontal* knowledge that is generated by the design team (Konda et al. 1992). This will include information regarding the process of design, examples of how design objects are decomposed, and knowledge that is specific to the design project, often at the interfaces of disciplines (Reddy et al. 1997). Separately, various service departments in the design organization will be gathering information from all projects. This could be called *historical* knowledge and is often quite structured in nature. The distinction between horizontal and historical knowledge can be seen as mapping on to that possessed by generalists and experts respectively (Baird et al 2000).

Product and service departments of organizations generate different kinds of information (Dias et al. 2002). The latter tend to produce generalizations based on historical data that is abstracted from the horizontal information created by the product departments. There is probably a special need today to document horizontal information, because it is unstructured and difficult to capture. On the other hand, it constitutes information in its most primitive form, and such process information can be invaluable to other product teams if captured and made available. An activity sequence diagram with an identified critical path can be seen as a piece of horizontal knowledge. Horizontal knowledge can also be seen as a collection of stories or narratives. It must be noted that the idea of individual stories (as opposed to 'grand' overarching theory or doctrine) is a central tenet in both existentialist and postmodernist philosophy, of which Heidegger is a key figure. The focus in such philosophy is not on overall unifying theory, but rather on the features of particular events. This can also be called a 'bottom up' approach to knowledge, as opposed to a 'top down' one.

8.5 Practice Based Knowledge Modelling: Examples

8.5.1 Modelling Tacit Knowledge: Construction Bid Decisions

Let us now look at some examples where AI techniques have been used to capture, structure and process practice based knowledge. Consider the modelling of tacit knowledge. There are many areas in engineering that are characterized by such knowledge, none more clearly than bidding for construction projects, decisions for which have been described as being made "on the basis of intuition derived from a mixture of gut feelings, experience and guesses" (Ahmad 1990)—the language is very reminiscent of Polanyi. There is also a wide acknowledgement of the poverty of theoretical approaches to this problem.

Hence, a backpropagation neural network called ANNBID was trained to make decisions on percentage mark-up for construction bids, based on the levels (i.e. numbers from 1 to 5) assigned to a set of 6 factors—i.e. nature of job, nature of client, location of project, risk involved in investment, current workload, and competition among contractors. The identification of these key factors was based on an industry-wide survey (Dias and Weerasinghe 1996). For the first three factors the scale ranged from 1 = easiest to 5 = most difficult, for the next two from 1 = lowest to 5 = highest, and for the last from 1 = highest to 5 = lowest. The scales were arranged such that a high score would tend to require a high mark-up. Data was obtained on 31 past cases from a single contractor, who assigned the above levels to all input factors and indicated the percentage mark-up he had used for each case. The network was trained on 27 of these cases, each of which had 6 input values (corresponding to the 6 factors) and a single output, corresponding to the mark-up, which ranged from 11% to 48%. The training consisted of generating a mapping between the inputs and the output for all 27 cases, such that the error between the network-generated and declared outputs was below a specified target value for all 27 cases. The remaining 4 cases were used to test the predictions of the trained network, which turned out to be quite good. The contractor now had a neural network that 'thought like he did' with respect to bid decisions; he could use the network in future bids to guide his guesses. Other researchers have also used such neural networks to model construction bid decisions (Hegazy and Moselhi 1994).

8.5.2 Modelling Shared Practice: Layout Design

The objective of this study was to explore the potential for using Artificial Neural Networks (ANNs) and Case Based Reasoning (CBR) for suggesting column spacing and sizing in multistorey buildings, based on historical examples (Dias and Padukka 2005). Column spacing and sizing are part of preliminary design, and often based on 'engineering judgement', which can be considered as partly dependent on shared practice. Data was obtained for a total of 45 existing buildings from different design offices; hence the data genuinely constituted shared practice, unlike in the tacit knowledge example, where it was a single contractor's 'gut feelings' that were modelled by the neural network.

ANNs have been described in the previous section. CBR is similar to ANN in the sense that it is also a connectionist approach. Each prior historical case in CBR can be conceptualized as an index card that contains the input and output values, the outputs being the column spacing and sizes that have been chosen in previous projects. If there is now a new design problem for which the inputs are known (as part of the design brief), a search is made among the cards for which the inputs had been within a specified percentage (say plus or minus 25%) of those pertaining to the new problem. This generates a subset of the prior projects, together with their overall degrees of similarity with respect to the new design inputs. The outputs of that subset are then averaged after being weighted with their similarity indices, to arrive at a prediction or recommendation for the desired design parameters of the new problem; if necessary the prediction can be based only on past cases that have similarity indices above a specified threshold (e.g. 70%).

For the column spacing problem, the inputs chosen were (i) type of building (residential/office); (ii) building height; (iii) type of foundation (pad/strip/raft/pile); (iv) type of slab (one-way/two-way); and (v) cost per unit area at Year 2000 prices. The output was the (minimum) column spacing. Training was carried out for the ANN on 34 of these cases, which were also used as the case base for CBR. The remaining 11 cases were used to test the ANN; and also as the 'new' cases in the CBR exercise. For the column sizing problem, the total number of cases was 29 (from among the above 45), with 21 being used for training and 8 for testing. The chosen inputs were (i) building height; (ii) tributary area; and (iii) concrete grade. The output was the column size, i.e. area, at basement (or ground) level. Two criteria were used to establish the success of the decision support tools, namely mean absolute error and the deviation from unity of the average ratio between predicted and desired outputs; these criteria were applied to the testing set.

In the column spacing exercise, the CBR results were better than the ANN ones on both criteria. After carrying out the ANN exercise, a sensitivity analysis was performed on the trained network, by evaluating the change in output when a given input is varied from its lowest value to its highest, all other inputs being held at their average values. This analysis revealed that building height and cost per unit area were the most significant inputs, with slab type being the next, and the others not so significant. In the column sizing problem too, the CBR results were slightly better than the ANN ones. At any rate, both ANN and CBR were found to be good AI tools for modelling shared practice. In other words, the trained ANN and CBR systems can be seen as having 'captured' the industry wide shared practice regarding column sizing and spacing; and can be used for suggesting those parameters for a new project, if the new design brief gives the required input values.

8.5.3 Cognitivist Modelling: Expert System for Cracks in Concrete

We now move from a connectionist to a cognitivist approach for structuring practice based knowledge. An expert system best reflects such an approach. Consider the system CONFAULT that seeks to diagnose the causes of cracking in concrete buildings (Dias 1992). The system first identifies what the major fault type is, from the six possibilities of **structural**, **movement**, **corrosion**, **durability**, **viscoelastic** and **early age**—note that **durability** refers to problems with the concrete itself, and **corrosion** to the specific durability problem of reinforcement corrosion. It does this through a set of 'meta rules' two of which are as follows:

> if cracked-element is *only secondary* AND crack-appearance is *oriented* AND age-of-cracking is *medium* AND development-of-cracking is *gradual*, then **movement** is VERY STRONGLY suggested.
>
> if cracked-element is *only secondary* AND crack-appearance is *oriented* AND age-of-cracking is *medium* OR development-of-cracking is *gradual*, then **movement** is STRONGLY suggested.

Here the hyphenated terms are the variables about which questions are asked; and the italicized ones their values, which can be selected out of a set of 2–3 pre-defined ones. The two rules above are required to cover the cases where (i) all 4 conditions are satisfied; and (ii) where 2 of the most important conditions and one other are satisfied.

Once the major fault type has been identified, the system selects a set of questions and rules to ascertain the specific fault. So for example we have the rules:

> if building-dimensions are *long*
>
> then building-geometry is *settle-critical* {0.6};
>
> if building-dimensions are *short*
>
> then building-geometry is *not-settle-critical* {0.6};

if building-loading is *uneven*
then building-geometry is *settle-critical* {0.8};
if building-loading is *even*
then building-geometry is *not-settle-critical* {0.6};
if building-stiffness is *low*
then building-geometry is *settle-critical* {0.9}

The numbers in parentheses are 'confidence factors' which can be used in assigning a level of confidence for belief in the rule. This is a reflection of uncertainty, which is an important feature of engineering reasoning (see Sect. 2.3). Interval probability approaches would use both upper and lower levels of confidence (Blockley 2013). The inference process can be several levels deep and hierarchically structured. The final diagnosis could be of the following form:

Movement STRONGLY suggested.
  Movement is *temperature* {0.57}
    not-settlement {0.47}
    settlement {0.17}
    not-temperature {0.13}

This type of result still requires the user to make a judgement, but the system gives assistance towards that end. For that reason, expert systems (and all other AI systems) are probably better termed 'decision support systems'. In expert systems of course the structure of the system and the rules have to be elicited from an expert (or from expert literature). This could be difficult, because experts may not always be able to articulate their practice based expertise precisely in the form of rules. Once a system is developed however, it will not only make suggestions for a new situation (i.e. an observed crack), but also divulge its line of reasoning. In other words, the system not only gives decision support but also explanations for the same.

8.5.4 Modelling Horizontal Knowledge: Vulnerability of Buildings to Bomb Blast

The examples given above are based essentially on historical knowledge that has been structured into various fields (i.e. factors that affect bid mark-up, grid spacing, and concrete cracking). Such structuring imposes a cognitive framework on the practice based knowledge and hence departs somewhat from the connectionist paradigm. The weighting and interactions between the factors however are genuinely connectionist in the first two examples, in that there are no cognitive rules that combine evidence in neural network type approaches. Horizontal knowledge (or knowledge as narrative) is less structured, and we shall now consider an example of that. The application area is the vulnerability of buildings to bomb blast. To be sure, there are numerical methods for solving blast load problems. However, where overall vulnerability (inclusive of human injury and death) is concerned, the problem is a socio-technical one, and vulnerability often depends more on non-structural factors (Chandratilake and Dias 2004).

In order to tackle this problem, a hierarchical causal tree was constructed by perusing the case descriptions of 10 blast events, having variations in the types of structures that were targeted, the nature of explosions, the physical and social context, and the intentions of terrorists. This was done using a Grounded Theory approach (Glaser and Strauss 1967). There were 63 'phenomena' extracted from the cases. Examples of such phenomena are "long and accurate warnings reduce human casualties"; "reinforced concrete framed construction can withstand considerable blast pressure" etc. By writing each phenomenon on a separate card, and by constant comparison among them, it was possible to cluster them appropriately and generate higher order 'concepts' that emerge near the top of the causal tree—e.g. 'physical entity', 'spatial planning', 'context' etc. Such clustering resulted in the insight that the two most important factors governing vulnerability were "advance warnings" and "the amount of glass used in buildings" (Chandratilake and Dias 2004).

After constructing the fault tree, it can be used in a semi-quantitative way to estimate either a numerical interval between 0 and 1 or a linguistic label (e.g. low, moderate, high) for the top level concept of vulnerability. This estimate is based on the linguistic labels (and associated levels of confidence in those labels) assigned by an assessor to the lowest level phenomena (Sanchez-Silva et al. 1995; Dias and Chandratilake 2005). The approach uses interval probability theory (Cui and Blockley 1990; Blockley 2013), itself based on fuzzy set theory. Another type of diagram that can be constructed out of event descriptions is what is called an event sequence diagram (Toft and Reynolds 1994). These depict sequential relationships between events that lead, for example to failure. Such diagrams have also been structured using connectionist type AI approaches (Stone et al. 1989).

Approaches such as grounded theory are 'faithful' to experience based data, and hence much more context based compared to theoretical solutions. The identification of phenomena and connections between them are of course done by the researchers, and this may introduce cognitive bias. Automation of such procedures, for example through co-word analysis (Monarch 2000) could be explored. For data to be converted to knowledge however, some structuring, and hence cognitivism, is almost always involved. This is true of all the knowledge structuring described above, at some stage of the structuring, e.g. in the definition of fields and factors, or the generation of concepts. Whether connectionist or cognitivist however, all the examples in Sects. 8.5.1–8.5.4 are ones where knowledge is derived from practice; they also present different ways in which practice based knowledge can be formalized (e.g. ANNs, CBR, expert systems and grounded theory).

8.6 Polanyi, Heidegger and Artificial Intelligence

At this stage we discuss an issue each from the philosophical and computational aspects in this chapter and seek to further clarify their inter-relatedness. The first has to do with a comparison of Polanyi and Heidegger. On the one hand, they are poles apart. Heidegger is a very nihilistic philosopher who advocated a 'hermeneutic of suspicion', while Polanyi sought to restore a fiduciary (or faith like) framework for the practice of science. Polanyi's focus is on epistemology, so he deals with the way that a human subject apprehends knowledge. Heidegger's focus is on ontology, and the notion of an individual human subject for him is a derivative (and even deficient) way of being, in a world that is characterized by shared practice and a network of relationships (both animate and inanimate). Both however focus on practice and it is this commonality that has resulted in their being thrown together in this chapter (see also Dias 2008). We could say that practitioner involvement is important to Polanyi and context dependence to Heidegger. It is interesting that both these aspects are foundational to an engineering approach, which relies heavily on practice based knowledge (in addition to theoretical knowledge).

Another apparent difference between Polanyi and Heidegger can be seen in their directions of movement where wholes and parts are concerned. For Polanyi, as described earlier (Sect. 8.2), knowing is associated with moving from parts to the whole. Knowledge of the whole is an emergent property. On the other hand, where Heidegger is concerned, the whole (or web of relations) is prior to any part thereof. A focus on any part is a derivative (or reductionist) move (see Sect. 7.2). But hierarchical structuring is common to both of them—bottom up in Polanyi's case and top down in Heidegger's. Networks and connectedness are important for them too. It is for this reason that AI approaches using a connectionist paradigm have been emphasized for reflecting their ideas and for modelling practice based knowledge.

This brings us to the second issue for discussion, namely the appropriateness of AI for reflecting the philosophical ideas of Polanyi and Heidegger on the one hand, and for modelling practice based knowledge on the other. It should be noted that the goal of Artificial Intelligence (AI) in general is the solving of practical problems, based very often on experience or heuristics. This places AI firmly within a paradigm of practice. The actual AI techniques themselves—even a highly connectionist AI technique such as an artificial neural network (ANN)—may in fact be highly computational and algorithmic. However, the inputs of an ANN are able to accommodate context related information and practitioner involvement—e.g. by converting qualitative information to a number scale. The outputs could be seen as mimicking practitioner judgement—e.g. the trained backpropagation network described in Sect. 8.5.1 behaves similarly to a human 'gut reaction' in arriving at bid mark-up decisions. They could also be seen as embodying shared practice—e.g. the trained backpropagation network and case base described in Sect. 8.5.2 encapsulate the way that designers across the industry choose column spacing and size.

It should also be noted that AI techniques within the cognitivist paradigm, such as rule based expert systems (Sect. 8.5.3), are also useful for practice based knowledge;

although they do not reflect too well the ideas of Polanyi and Heidegger. In fact, as shown in Sects. 8.2 and 8.3, both of them, either implicitly (Polanyi) or explicitly (Heidegger) rejected in principle the validity of cognitive modelling. Whether connectionist or cognitivist however, AI gives us a way for formalizing practice based knowledge at a technical or detailed level. We saw earlier in Sect. 2.5 that systems thinking can help us to formalize practice at a conceptual or broad level. Taken together then, systems thinking at a conceptual level and AI at a technical level can be seen as ways to formalize practice based knowledge; and through that to elevate its intellectual status. In some ways, this is reminiscent of Herbert Simon's (1996) call for search strategies or decision sciences to be emphasized in engineering programs at universities; he called them 'sciences of the artificial'.

8.7 Summary

- The epistemology of Michael Polanyi and the ontology of Martin Heidegger provide a compelling intellectual basis for the notion of practice based knowledge.
- Artificial Intelligence (AI) techniques such as Artificial Neural Networks (ANN) and Case Based Reasoning (CBR) can model philosophical concepts such as tacit knowing (Polanyi) and shared practice (Heidegger).
- The above philosophical grounding and computational formalizations for practice based knowledge have highlighted the conceptual differences between connectionist and cognitivist approaches within AI. Both approaches are useful for modelling practice based knowledge. The connectionist approach is more reflective of the ideas of Polanyi and Heidegger; however, all attempts to generate knowledge from data would require a degree of cognitivism.
- Examples are given of how historical practice based knowledge can be captured, structured and processed using AI approaches such as ANN, CBR and Expert Systems; and horizontal knowledge using Grounded Theory combined with Interval Probability Theory, itself based on fuzzy set theory.

Acknowledgements Adapted from *Knowledge Based Systems, 20*(4), 382–387, Philosophical grounding and computational formalization for practice based engineering knowledge by W. P. S. Dias, 2007, with permission from Elsevier.

References

I. Ahmad, Decision support system for modelling bid/no-bid decision problem. ASCE J. Constr. Eng. Manag. **116**(4), 595–608 (1990)

F. Baird, C.J. Moore, A.P. Jagodzinski, An ethnographic study of engineering design teams at Rolls Royce Aerospace. Des. Stud. **21**(4), 333–355 (2000)

D.I. Blockley, Analysing uncertainties: towards comparing Bayesian and interval probabilities. Mech. Syst. Signal Process. **37**(1–2), 30–42 (2013)
R. Chandratilake, W.P.S. Dias, Identifying vulnerability of buildings to blast events using Grounded Theory, in *Proceedings of the 10th Annual Symposium on Research for Industry*, Engineering Research Unit, University of Moratuwa, Moratuwa, Sri Lanka (2004)
R.D. Coyne, Design reasoning without explanations. AI Mag. **11**(4), 72–80 (1990)
W.C. Cui, D.I. Blockley, Interval probability theory for evidential support. Int. J. Intell. Syst. **5**(2), 183–192 (1990)
W.P.S. Dias, CONFAULT—an expert system for fault diagnosis in reinforced concrete structures. Civ. Eng. Syst. **9**, 147–160 (1992)
W.P.S. Dias, Reflective practice, artificial intelligence and engineering design: common trends and inter-relationships. Artif. Intell. Eng. Des., Anal. Manuf. (AIEDAM) **16**, 261–271 (2002)
W.P.S. Dias, Heidegger's resonance with engineering: the primacy of practice. Sci. Eng. Ethics **12**(3), 523–532 (2006)
W.P.S. Dias, Philosophical grounding and computational formalization for practice based engineering knowledge. Knowl. Based Syst. **20**(4), 382–387 (2007)
W.P.S. Dias, Philosophical underpinning for systems thinking. Interdisc. Sci. Rev. **33**(3), 202–213 (2008)
W.P.S. Dias, D.I. Blockley, Reflective practice in engineering design. ICE Proc. Civ. Eng. **108**(4), 160–168 (1995)
W.P.S. Dias, S.R. Chandratilake, Assessing vulnerability of buildings to blast using Interval Probability Theory, in *Proceedings of the 8th International Conference on the Application of Artificial Intelligence to Civil, Structural and Environmental Engineering*, ed. by B.H.V. Topping (Civil-Comp Press, Stirling, 2005)
W.P.S. Dias, U.A. Padukka, AI techniques for preliminary design decisions on column spacing and sizing, in *Proceedings of the 8th International Conference on the Application of Artificial Intelligence to Civil, Structural and Environmental Engineering*, ed. by B.H.V. Topping (Civil-Comp Press, Stirling, 2005)
W.P.S. Dias, R.L.D. Weerasinghe, Artificial neural networks for construction bid decisions. Civ. Eng. Syst. **13**, 239–253 (1996)
W.P.S. Dias, E. Subrahmanian, I.A. Monarch, Dimensions of order in engineering design organizations. Des. Stud. **24**, 357–373 (2002)
H.L. Dreyfus, Husserl, Heidegger and modern existentialism, in *Great Philosophers: An Introduction to Western Philosophy*, ed. by B. Magee (Oxford University Press, Oxford, 1988), pp. 252–277
B. Glaser, A.L. Strauss, *The Discovery of Grounded Theory: Strategies for Qualitative Research* (Weidenfeld and Nicolson, London, 1967)
F. Hayes-Roth, D.A. Waterman, D.B. Lenat (eds.), *Building Expert Systems* (Addison-Wesley, London, 1983)
T. Hegazy, O. Moselhi, Analogy-based solution to markup estimation problem. ASCE J. Comput. Civ. Eng. **8**(1), 72–87 (1994)
M. Heidegger, *The Question Concerning Technology and Other Essays* (trans: W. Lovitt) (Harper & Row, New York, 1977)
M. Heidegger, *Being and Time* (trans: J. Stambaugh) (SUNY Press, Albany, 1997)
M. Johnson, *The Body in the Mind: The Bodily Basis of Meaning, Imagination and Reason* (University of Chicago Press, Chicago, 1987)
M. Johnson, *The Meaning of the Body* (University of Chicago Press, Chicago, 2007)
S. Konda, I. Monarch, P. Sargent, E. Subrahmanian, Shared memory in design: a unifying theme for research and practice. Res. Eng. Design **4**(1), 23–42 (1992)
A. Koestler, *The Ghost in the Machine* (Picador, London, 1967)
M. Minsky, Logical vs. analogical or symbolic vs. connectionist or neat vs. scruffy. AI Mag. **12**(2), 34–51 (1991)

I.A. Monarch, Information science and information systems: converging or diverging? in *CAIS 2000, Dimensions of a Global Information Science, Proceedings of the 28th Annual Conference.* Canadian Association for Information Science (2000)
M. Polanyi, *Personal Knowledge: Towards a Post-critical Philosophy* (University of Chicago Press, Chicago, 1958)
M. Polanyi, *The Tacit Dimension* (Doubleday & Co., Garden City, 1966)
M. Polanyi, in *Knowing and Being*, ed. by M. Greene (University of Chicago Press, Chicago, 1969)
H. Prosch, *Michael Polanyi: A Critical Exposition* (SUNY Press, Albany, 1986)
J.C. Reddy, S. Finger, S.L. Konda, E. Subrahmanian, Designing as building and re-using of artifact theories: understanding and support of design knowledge. in *Proceedings of the Workshop on Engineering Design Debate* (University of Glasgow, Glasgow, 1997)
M. Sanchez-Silva, C.A. Taylor, D.I. Blockley, Evaluation of proneness to earthquake-induced failure of buildings in Buenaventura, Colombia, in *Structures to Withstand Disaster*, ed. by D. Key (Thomas Telford, London, 1995), pp. 137–152
D.A. Schon, *The Reflective Practitioner: How Professionals Think in Action* (Temple Smith, London, 1983)
H.A. Simon, *The Sciences of the Artificial*, 3rd edn. (MIT Press, Cambridge, 1996)
J.R. Stone, D.I. Blockley, B.W. Pilsworth, Towards machine learning from case histories. Civ. Eng. Syst. **6**, 129–135 (1989)
B. Toft, S. Reynolds, *Learning from Disasters* (Butterworth-Heinemann, Oxford, 1994)

Chapter 9
Conclusion: From Philosophy to Engineering

9.1 Looking Back

9.1.1 The Path We Have Taken

We started in Chap. 1 by highlighting some things that are crucial for engineering—practice, context, ethics, models and failure; and discovered that four eminent 20th century philosophers had very relevant insights concerning them. After establishing the relevance of ethics, ontology and epistemology for engineering in Chap. 2, we covered the contributions of the philosophers with a chapter each for Karl Popper, Thomas Kuhn and Michael Polanyi; and two for Martin Heidegger. Another chapter drew on both Polanyi and Heidegger. Chapter 1 allocated the various issues to the different philosophers we turned to. However, in this concluding chapter we look at commonalities across all the philosophers on each of these issues.

9.1.2 Practice

Heidegger is the main proponent of practice. His view is that we know about the world by absorbing 'pre-theoretical shared agreement in practices'. Theoretical knowledge comes later—more often than not when there are breakdowns in everyday activity, as experienced by the carpenter he described (Sect. 7.2). Theoretical knowledge can be sought deliberately too, when someone seeks to escape from 'average everydayness' in order to seek 'authenticity' (Sect. 7.5). So for engineers who are sometimes labelled as merely 'makers' rather than 'thinkers'—or *homo faber* rather than *homo sapiens*—Heidegger's primacy of practice provides a philosophical underpinning for their stature. We saw in Chap. 2 that engineers are scientists and theoreticians too, but much more than that (not less) when they act as managers and practitioners respectively. Polanyi adds his weight to the importance of practice, by pointing to both the heuristic passion and sober judgement (as opposed to 'cool detachment') that is required of

P. Dias, *Philosophy for Engineering*, SpringerBriefs in Applied Sciences and Technology, https://doi.org/10.1007/978-981-15-1271-1_9

scientists in their pursuits (Sects. 5.2 and 5.5). He also argues that knowing involves doing, citing the example of apprentices (whether cabinet makers or scientists) who train under supervisors—because the latter 'know more than they can tell' and can pass on such 'tacit knowledge' only by having others practise under their guidance (Sect. 8.2).

9.1.3 Context

Heidegger underlines the importance of context too. He describes a 'web of relations' that we are a part of—a web that cannot properly be 're-presented' by models of it (Sect. 8.3). He also talks about humans being 'thrown' into various contexts that he describes as 'situations' (Sect. 7.3 and Table 7.1). It is such 'thrownness' in a 'situation' that calls for decisive action and helps us to realize our 'authenticity' (Sect. 7.3). This emphasizes the importance of practice as well; and all engineers working in projects with tight time and budget constraints, while pitting their wits against nature as well as culture, will identify with 'thrownness'.

Although the idea of context is not specifically referred to by Kuhn, we get from him the notion that theories about the world can differ because they are human constructs. In applying this to engineering, we have said that models need to be dependable by aiming for completeness, rather than accurate by aiming for simplicity (and ignoring many aspects of the world)—Sect. 4.5 and Table 4.2. In other words, we are saying that context is crucial for the practice of engineering, as opposed to decontextualized generalizations that are more the goal of science. It is context also that provides constraints, which help (or force) us to 'tailor' our designs or practice creatively.

9.1.4 Ethics

There two main aspects of ethics we have come across, once again presented mostly by Heidegger and Polanyi. The first is the question of moral responsibility, which Polanyi charges individual scientists to hold, in conjunction with their intellectual freedom, in the process of propounding theories (Sect. 5.5). For engineers, we saw that their primary ethical responsibility was not to their clients or employers, but rather to the safety of the public. We also saw how professional engineering institutions are meant to safeguard professional ethics through their codes of conduct (Sect. 5.7); this parallels the way that the scientific community is meant to regulate the production of scientific knowledge through peer review (Sect. 5.5).

For Heidegger our responsibility is to develop a 'hermeneutic of suspicion', especially with respect to technology—for example not to be taken in by the idea that technology is neutral (Sect. 6.2). We argued in Sect. 6.4 that technology is not neutral at various levels of subtlety; and needs to be judged by the way it promotes or diminishes safety, justice, community and humanness. The term 'care' is also associated with Heidegger, and we saw how engineering could be viewed as having an 'ethic of care', as the medical profession is—in other words, it is a 'duty of care' that engineers need to exercise towards others (Sect. 6.5). Popper too touches on ethics through his discourse on an 'open society'. Drawing parallels with his ideas about the asymmetry between verification and falsification (Sect. 3.2), he argues that social goals should not be aimed at maximizing happiness but rather on minimizing unhappiness (Sect. 5.6). In addition, while critiquing all social utopian goals based on social trends (which are not really laws), Popper's prescription for social change is 'piecemeal social engineering' (Sect. 3.3).

The other aspect of ethics is the pursuit of beauty or aesthetics. We saw this in Polanyi, who not only advocates for beauty and elegance in scientific theories, but also describes scientists as doing so anyway—thus being both descriptive and prescriptive about it (Sect. 5.2). We presented an example from structural engineering too that endorsed the well know maxim that "if it looks right, it *is* right" (Sect. 5.4). Heidegger's contribution to beauty is to critique the reductionism and rationalism inherent in modern technology by comparing it to more holistic forms of 'bringing forth'—notably poetry (Sect. 6.3). His ideas have been used to suggest metaphor rather than method as a fruitful process for creative design (Sect. 6.5).

9.1.5 Models

Both Popper and Kuhn deal extensively with theories or paradigms—i.e. representations of the world. Popper privileges theory over observation; it is important to start with a bold theory, which can later be checked via observation (Sect. 3.2). Engineers represent the world in models of it. They model their proposed solutions or designs for checking against the required specifications using critical tests. When their designs are fabricated, it is the 'calculation procedure model' used for their designs that is critiqued by its performance in the world. So engineers recognize that models are useful, but have to be improved (Sects. 3.4 and 3.5).

Kuhn is more critical of theories, saying that they are merely social constructs (Sect. 4.3). Engineers recognize this too of their models; but also the truism that while all models are wrong, some are indeed useful. So engineers aim for models that are appropriate and as complete as possible, and dependable for delivering safety. Although completeness is what engineers aim for, they recognize that they have to contend with the three aspects of uncertainty, namely randomness, fuzziness and incompleteness (Sect. 2.3). Structural engineering has been described as "the art of moulding materials we don't wholly understand, into shapes we can't fully analyze, so as to withstand forces we can't really assess, in such a way that the community at large has no reason to suspect the extent of our ignorance" (attributed to Dr. A. R. Dykes at 1946 Chairman's Address to the Scottish Branch of the Institution of Structural Engineers).

There are ways for dealing with these elements of uncertainty (Sect. 2.3); but engineers, perhaps in the spirit of Heidegger, should remain suspicious of them. Heidegger's own contribution to the concept of a model is that the world in all its richness cannot be fully captured by any representation of it (Sect. 8.3). One specific area that merits suspicion is the increasingly powerful software that is available for modelling engineering response. Power and precision do not necessarily deliver accuracy and correspondence with the world; and engineers need to make allowance for that through a variety of checks and balances. They also need to pay adequate attention to 'things that count but cannot be counted'.

9.1.6 Failure

Popper is the main exponent of failure; he advocates that scientists should try to falsify their own theories and that of others—because that is the way scientific knowledge grows (Sect. 3.2). In our application of his ideas to engineering, we saw that engineers do the same to their proposed solutions in the design phase. However, genuine testing of and improvements to the 'calculation procedure model' take place only when the realized designs are tested by the world. This occurs when there are real world engineering failures, but we cannot plan for them to happen; this is why engineering knowledge grows more slowly than its scientific counterpart (Sect. 3.5). When failures do take place, we can try to isolate the specific component of engineering knowledge that has failed, so that changes can be made thereto (Sects. 3.5 and 3.6). Heidegger uses the term 'breakdown' to describe a disjoint in the everyday functioning of our world (Sect. 7.2). Breakdowns can force us to study the fundamental nature of things that make up such everyday functioning. This could range from the properties of physical entities, through the social entities (e.g. individuals and groups) that use them, to the connections between such differing entities.

9.2 Looking Forward

9.2.1 Paths not Trodden

It is not claimed that this book has covered all aspects that are crucial to engineering. For example, although we have given some coverage to *uncertainty* in Sects. 2.3 and 8.5.3, we have not majored on it. We have not given much treatment to the notion of *sensitivity* either, although mentioned briefly in Sect. 8.5.2 (but see Dias 1997). *Hierarchical structuring* and *emergence* are also important for engineering, covered by both Polanyi (Sect. 8.2) and Simon (1996, pp. 188–190), notably through his tale of the two watchmakers of antiquity. Engineering *approximation*, touched on briefly in Sects. 2.3, 3.7 and 4.5, is another area worth exploring. This is the way that engineers make complexity tractable—whether through idealization or discretization (e.g. in the finite element method that is used widely in all branches of engineering). Finally, although *design* is central to engineering activity, we have not treated it separately, because practice, context, ethics, models and failure are all related to or impinge on engineering design, which is referred to in many parts of the book. Mention should be made in particular of Goldman's (2017) assertion that the rationality of engineering is best reflected in design.

While this book has presented engineering as being *different* from science (Fig. 2.1 and Sect. 4.5), we have acknowledged the *centrality* of engineering science to engineering knowledge (Fig. 2.2). We have also referred to the *complementarity* of scientific and engineering approaches—note the relationships between reflective practice and technical rationality (Sects. 2.4 and 8.2); and between knowing how and knowing what (Sects. 2.4 and 7.2). We have argued however that engineering both encompasses and is more sophisticated than science (Fig. 2.2 and Sects. 2.3, 7.2). What we have not done is to focus on the actual interplay between science and engineering, and more could be done in this regard (e.g. Dias and Blockley 1995). This could lead to clarity on what disciplines should bear the 'engineering' label. Can we for example have a discipline called 'financial engineering'—given that it is not rooted in the natural sciences? It is nature that provides the constraints and opportunities for engineering to engage with. The original definition for civil engineering (at that time covering all branches of engineering other than military engineering) in the Royal Charter of 1828 for the Institution of Civil Engineers, U.K. (the world's first professional engineering body) was "the art of directing the great sources of power in nature for the use and convenience of man", and featured nature strongly.

In some ways, we may have been limited because we confined ourselves to engaging with our four philosophers. Three of them are clearly known as philosophers of science. Since engineering is both an art and science, there is scope for reflecting on engineering through the lens of the arts too; and also the social sciences. The present author has compared engineering with history, because the particularities of context are important for both those disciplines (Dias 2014). Legal design has been likened to engineering design (Howarth 2013). Justice is another theme around which engineering can be discussed, since engineering is embedded in a social matrix; we have dealt with it briefly in Sect. 6.4.

It is not very fashionable to consider the influence of religion on various phenomena, probably because it tends to be divisive. However, religion has been foundational in both the pursuit of inquiry (almost all ancient universities started off as scholastic monasteries); and also the practice of engineering, most notably that of structural engineering, in most parts of the world—consider the way that cathedrals, mosques, temples and stupas are central to tourism today. In addition, religion has been a major factor in shaping both societal culture and worldview—e.g. the way people think about the world and their interventions in it. Furthermore, there is a considerable body of literature on the relationship between religion and science to which the present author has also made a modest contribution (Dias 2010), interestingly through a Polanyian perspective. Kuhn too pointed out that some aspects of normal science had similarities to orthodox theology (Sect. 4.2). Polanyi was trying to recover a fiduciary mindset for the practice of science—something he felt had been lost through the critical movement. His major work is subtitled *Towards a post-critical philosophy* and contains this very poignant passage (Polanyi 1958, pp. 265–6):

> The critical movement, which seems to be nearing the end of its course today, was perhaps the most fruitful effort ever sustained by the human mind. The past four or five centuries, which have gradually destroyed or overshadowed the whole medieval cosmos, have enriched us mentally and morally to an extent unrivalled by any period of similar duration. But its incandescence has fed on the combustion of the Christian heritage in the oxygen of Greek rationalism, and when this fuel was exhausted the critical framework itself burnt away.

Heidegger on the other hand was a very nihilistic philosopher associated with the 'death of God' movement of Nietzsche, and advocated a 'hermeneutic of suspicion', as we have seen (Sect. 6.2). Nevertheless, we also saw in Sect. 8.6 that there is some remarkable similarity or at least complementarity between some of his ideas and those of Polanyi. So, while there is considerable scholarship on the interactions among religion, philosophy and science, very little has been done to bring engineering into this discourse. Note also the importance of *trust* for the engineering process, which is an ethical issue (Sect. 5.7), and often associated with religion. Hence the nexus between engineering and religion is well worth pursuing.

9.2.2 Where Do We Go from Here?

It is hoped that this book would stimulate the teaching of a discipline such as the philosophy of engineering within engineering programs. Many accreditation bodies worldwide now require some humanities credits within such programs. A course on philosophy made relevant to engineering could be very appropriate. Approaches taken in such courses may be very different to that taken in this book. The point however is that the practice of engineering would be well served by some formation of practitioners in ways to think reflectively and philosophically about it. Such courses could be taught by either by engineers or philosophers, or both; but it may require humanities departments or faculties to nurture them.

It would also be desirable if engineering departments think about the formalization of practice based knowledge. This is also an area that is open to a diversity of approaches. This book has suggested systems thinking approaches (Sect. 2.5) and Artificial Intelligence (AI) techniques (Chap. 8) as two possibilities. There are doubtless many more. Herbert Simon, the Nobel prize winning economist, psychologist and computer scientist, also acknowledged as one of the four founding fathers of AI, suggests in his book *The Sciences of the Artificial*, that engineering programs be populated by courses on design theory and theories of search; rather than focusing on properties of materials, which are better taught in faculties of science (Simon 1996, Chap. 5, pp. 111–138).

Finally, it is hoped that this book would have stimulated at least a few engineers or engineering students to read the four philosophers for themselves. The works of

Karl Popper and Martin Heidegger tend to be voluminous and are best accessed via their selected works (Popper 1983; Heidegger 1977). Thomas Kuhn's major work is short enough to read (Kuhn 1970); while the best way to start on Michael Polanyi is through his book on tacit knowing (Polanyi 1966). Heidegger is the most difficult to read; but if engineers get their heads around Heidegger, he could well become their patron philosopher.

9.3 Summary

- Both Heidegger and Polanyi have insights regarding practice and ethics. In addition, Heidegger, by making contributions regarding context, models and failure as well, could well qualify to be the patron philosopher for engineers.
- Both Popper and Kuhn have insights regarding models, while Popper has contributions on ethics and failure too; and Kuhn on context. So next to Heidegger, Popper is probably the most important philosopher for engineers.
- Other characteristics of engineering not explored much in this book are uncertainty, sensitivity, hierarchy, emergence and approximation. In addition, although we have considered the relationship between science and engineering, the actual interplay between them could be studied further. The notion of design is considered to be central to engineering rationality, and is impinged on by all the characteristics of practice, context, ethics, models and failure that we have explored in depth.
- Engaging with the arts, social sciences and religion could also be fruitful to obtain other perspectives on engineering; in the way that we have engaged with science.
- Philosophy of engineering and practice based knowledge are commended as two disciplines in their own right that could extend the ideas presented in this book.

References

P. Dias, Is science very different from religion? A Polanyian perspective. Sci. Christ. Belief **22**(1), 43–55 (2010)

P. Dias, The disciplines of engineering and history: some common ground. Sci. Eng. Ethics **20**(2), 539–549 (2014)

W.P.S. Dias, Sensitivity and substitutability in concrete construction. Asia Pac. Build. Constr. Manag. J. **2**(2) (1997)

W.P.S. Dias, D.I. Blockley, Reflective practice in engineering design. ICE Proc. Civ. Eng. **108**(4), 160–168 (1995)

S.L. Goldman, Compromised exactness and the rationality of engineering (Chap. 1), in *Social Systems Engineering; The Design of Complexity*, ed. by C. Garcia-Diaz, C. Olaya (Wiley, Oxford, 2017), pp. 13–29

M. Heidegger, in *Basic Writings*, ed. by D.F. Krell (Harper Collins, New York, 1977)

D. Howarth, *Law as Engineering: Thinking About What Lawyers Do* (Edward Elgar, Cheltenham, 2013)

T.S. Kuhn, *The Structure of Scientific Revolutions*, 2nd edn. (University of Chicago Press, Chicago, 1970)
M. Polanyi, *Personal Knowledge: Towards a Post-critical Philosophy* (University of Chicago Press, Chicago, 1958)
M. Polanyi, *The Tacit Dimension* (Doubleday & Co., Garden City, 1966)
K.R. Popper, in *Popper Selections*, ed. by D.W. Miller (Princeton University Press, Princeton, 1983)
H.A. Simon, *The Sciences of the Artificial*, 3rd edn. (MIT Press, Cambridge, 1996)